Foundations of Numerical Methods and Data Analysis

Foundations of Numerical Methods and Data Analysis

Tanvir Mustafy
and
Md. Tauhid Ur Rahman

CWP

This edition has been published by Central West Publishing PTY LTD, (ABN 13 683 898 722) Australia

For more information about the books published by Central West Publishing PTY LTD, please visit https://centralwestpublishing.com

Disclaimer

Every effort has been made by the publisher, editors and authors while preparing this book, however, no warranties are made regarding the accuracy and completeness of the content. The publisher, editors and authors disclaim without any limitation all warranties as well as any implied warranties about sales, along with fitness of the content for a particular purpose. Citation of any website and other information sources does not mean any endorsement from the publisher, editors and authors. For ascertaining the suitability of the contents contained herein for a particular lab or commercial use, consultation with the subject expert is needed. In addition, while using the information and methods contained herein, the practitioners and researchers need to be mindful for their own safety, along with the safety of others, including the professional parties and premises for whom they have professional responsibility. To the fullest extent of law, the publisher, editors and authors are not liable in all circumstances (special, incidental, and consequential) for any injury and/or damage to persons and property, along with any potential loss of profit and other commercial damages due to the use of any methods, products, guidelines, procedures contained in the material herein.

NATIONAL LIBRARY OF AUSTRALIA

A catalogue record for this book is available from the National Library of Australia

ISBN (print): 978-1-922617-68-2

Preface

Numerical methods have always fascinated me because they sit right at the crossroads of mathematics, engineering, and computation. The field itself is not new, its foundations stretch back centuries, but its importance has only grown in the age of digital computing. Where once engineers and scientists relied on tedious manual calculations, today we harness the speed and accuracy of modern machines. Yet, despite the power of computers, the need for clear thinking and sound methodology remains unchanged.

This book was born from years of teaching, research, and practical engagement with real engineering problems. Time and again, we have seen students and professionals alike struggle not because they lacked computational tools, but because they lacked a structured pathway that connects mathematical theory with applied problem solving. Our intention here is to provide such a pathway. Alongside rigorous explanations, we have included examples, illustrations, and exercises that reflect the challenges one might encounter in practice. Our hope is that readers will find in these pages both a foundation and a springboard something that supports learning and also inspires exploration.

About the Authors

Dr. Tanvir Mustafy is a distinguished scholar, educator, and thought leader whose career bridges the frontiers of civil engineering, computational mechanics, and applied data science. His academic path has been both rigorous and global: a B.Sc. in Civil Engineering from BUET, an M.Sc. in Structural Engineering from the University of Alberta, and a Ph.D. in Computational Mechanics from the Université de Montréal, where he was awarded the Gold Medal for outstanding research contributions. To complement his scientific pursuits with managerial insight, he earned an MBA from the Institute of Business Administration (IBA), University of Dhaka, adding strategic and leadership depth to his technical expertise. Dr. Mustafy's professional footprint spans continents and disciplines. He has conducted postdoctoral research at Imperial College London, collaborated with leading research chairs in Canada, and served as visiting faculty at the Royal Military College of Canada and Aix-Marseille University in France. Today, he serves as Associate Professor in the Department of Civil and Environmental Engineering at North South University, Bangladesh, after previously holding the same position at the Military Institute of Science and Technology (MIST). A prolific researcher and author of over 100 journal papers and conference publications, Dr. Mustafy's work covers machine learning in structural engineering, earthquake resilience, advanced finite element modeling, structural dynamics, injury biomechanics, and sustainable infrastructure systems. His research has not only advanced theory but also shaped practice contributing to projects on flood early-warning systems, recycled water usage, and sustainable urban design solutions such as green roofs in tropical climates. Equally committed to teaching, mentorship, and professional service, he has guided the next generation of engineers with the same rigor and creativity that define his own work. Recognized for his contributions, he holds the title of Professional Structural Engineer (SEng) and is actively involved with professional bodies in Bangladesh and abroad. With a rare combination of intellectual depth, global exposure, and practical vision, Dr. Tanvir Mustafy continues to stand at the forefront of civil engineering research and education, shaping both knowledge and practice for a more resilient and sustainable future.

Dr. Tauhid Rahman finished his Ph.D. in Environmental Engineering from Tohoku University, Japan, in 2009. He did his M.Sc. in Environmental Engineering, Land, and Water Engineering from KTH, Sweden, and his B.Sc. in Civil Engineering from Bangladesh University of Engineering and Technology, Bangladesh. He is currently working as a professor in the CE Department of MIST. His research interests are water quality modeling, land use change detection, climate change, water insecurity, micro-climate effect, etc.

Table of Contents

CHAPTER 1

Introduction

1.1 Overview

Numerical methods involve formulating mathematical problems in a way that allows them to be solved using arithmetic operations. There are various numerical methods, all of which share a common characteristic: they require a significant amount of repetitive arithmetic calculations. The rise of fast and efficient digital computers has significantly increased the importance of numerical methods in engineering problem solving in recent years.

This book provides a thorough examination of numerical methods, combining theoretical principles with real-world examples. The course starts by covering fundamental subjects such as calculus, vectors, matrices, and programming basics. It then investigates into a detailed examination of error analysis and computer arithmetic. The book provides an overview of important methods and tools, including Horner's Rule and Euler's Method, and then looks into the numerical solutions of ordinary and partial differential equations. The text provides a comprehensive discussion of techniques for curve fitting, polynomial interpolation, and numerical integration, equipping readers with the necessary skills to effectively approximate solutions. This book provides comprehensive coverage of advanced topics such as solving systems of equations, the finite element method, and spectral methods. It serves as a valuable resource for gaining a deep understanding of numerical methods and their practical applications in mathematical and engineering contexts. This book offers a valuable learning resource for individuals seeking to develop expertise in operating system functions. The topics have been organised in a logical manner, with the inclusion of relevant examples and illustrations to enhance understanding and facilitate effective recapitulation.

1.2 Historical and Bibliographical Remarks

The development of numerical methods is closely connected to the progression of mathematics and computational science. The methods discussed in this book have a rich history that dates back centuries. Mathematicians have been working on these methods to find practical solutions to complex problems long before the advent of modern computers. The origins of numerical analysis can be traced back to the 17th century when calculus was pioneered by Sir Isaac Newton and Gottfried Wilhelm Leibniz. Newton's contributions to interpolation and his method for solving systems of linear equations, which is now commonly referred to as Gaussian elimination after Carl Friedrich Gauss, represent significant milestones in the development of numerical methods. Leonhard Euler expanded upon these concepts and introduced the Euler's Method in the 18th century. This method revolutionised the field of solving ordinary differential equations (ODEs) by providing a systematic approach to numerical approximation.

Understanding the limitations of approximations in numerical computations has highlighted the significance of error analysis. Carl Friedrich Gauss's contributions in the 19th century regarding least squares and error propagation have been instrumental in shaping modern error analysis. In the 20th century, the advancements in floating-point arithmetic and the formalisation of concepts like machine precision and round-off error by influential figures such as John von Neumann and Alan Turing greatly contributed to the dependability of numerical computations. The historical development of numerical solutions for both ordinary and partial differential equations is extensive. The Taylor series method, a key technique for approximating solutions, is based on the contributions of Brook Taylor and Joseph-Louis Lagrange. The Runge-Kutta methods, devised by Carl Runge and Martin Kutta in the early 20th century, were a notable breakthrough, providing improved precision and stability when solving ODEs.

PDEs, commonly found in disciplines like physics and engineering, were initially tackled through analytical techniques. Nevertheless, the shortcomings of these approaches prompted the emergence of numerical methods, specifically finite difference methods, which were established in the early 20th century by renowned mathematicians Richard Courant and Kurt Friedrichs. The study of curve fitting and interpolation dates back to ancient civilisations, where linear and polynomial approximations were employed in fields such as astronomy and engineering. Gauss formalised the least squares method, which later became a cornerstone of statistical regression analysis. Polynomial interpolation methods, pioneered by Joseph-Louis Lagrange and Isaac Newton, have proven to be invaluable in estimating unknown values using known data points.

The techniques used in numerical integration, such as the trapezoidal rule and Simpson's rule, can be traced back to the contributions of mathematicians like Thomas Simpson and James Gregory. The Newton-Cotes quadrature formulas were developed by mathematicians to improve the accuracy of computing definite integrals in cases where analytical solutions were not possible. The finite element method (FEM) is a significant technique in modern computational mechanics that was developed in the mid-20th century by engineers such as Richard Courant and John Argyris. The framework offered a solution for tackling intricate boundary value problems in the field of engineering. Spectral methods, which were developed later, provided an alternative approach to solving differential equations with a high level of accuracy, especially in cases where the solutions are smooth. The development of computers brought about a significant transformation in numerical analysis, enabling the execution of intricate algorithms that were previously impractical to perform manually. Programming languages like FORTRAN, which emerged in the 1950s, were specifically created to facilitate numerical computation and played a vital role in spreading numerical methods. Currently, modern numerical software packages such as MATLAB and NumPy are constantly advancing, integrating the most recent findings in numerical analysis and offering user-friendly resources for engineers and scientists. This concise historical overview highlights the ongoing development of numerical methods, motivated by the necessity to address progressively intricate problems in the fields of science, engineering, and mathematics. The methods discussed in this book have been refined over centuries of intellectual work and are still being improved to address the challenges of the modern computational landscape.

1.3 The Significance of Numerical Analysis in Various Fields

Numerical methods involve formulating mathematical problems in a way that allows them to be solved using arithmetic operations. Various types of numerical methods share a common characteristic: they require a significant amount of lengthy arithmetic calculations. In addition to numerical methods, there are two other approaches to problem solving: analytical methods and graphical methods. The latter is not effective because it lacks precision, as graphs are drawn by hand. This also makes the process long and laborious. Additionally, a drawback of this approach is its limited applicability to problems with three or fewer dimensions. Analytical methods provide valuable insights into the behaviour and characteristics of a specific system. Regrettably, these models are only applicable to systems that can be approximated using linear models. Given the intricate and nonlinear nature of practical problems, relying solely on analytical methods proves insufficient. In the past, analytical and graphical methods were favoured because of the time-consuming nature of numerical methods mentioned earlier. Performing various arithmetic operations was time-consuming due to the limited tools available, such as slide rules, calculators, and paper and pencil. The increased number of operations also heightened the risk of an incorrect outcome resulting from human error. The advent and widespread availability of affordable computers in recent times has revolutionised the understanding of numerical methods and transformed problem-solving approaches. A computer has the capability to execute an enormous number of calculations within a single second. Therefore, the issue of speed limitation related to numerical methods has been resolved.

Numerical methods play a crucial role in engineering as they provide a comprehensive understanding of various systems across different engineering disciplines. This is achievable because it focusses on the appropriate stages of problem solving. In engineering practice, problem solving can be broken down into three stages: Formulation, solution, and interpretation. Analytical methods primarily concentrate on the solution stage, aiming to develop a simplified model to provide an answer. On the other hand, numerical methods focus on the formulation and interpretation stages. In the formulation stage, one must establish the optimal relationship between the problem's dimensions and the fundamental laws of nature. Given the transformative impact of computing power, the formulation stage has gained significant importance as it plays a crucial role in generating the most optimal solution. Numerical methods also offer expanded possibilities during the interpretation stage. The mathematical model setup in the formulation stage can be easily adjusted multiple times to facilitate calculations and gain a deeper understanding of the system's characteristics.

There are various types of these modifications:

Systems of a larger scale: Additional dimensions/components have been incorporated into the mathematical model. This enhances the realism of the system.

Non ideal systems: Typically, systems are solved assuming ideal conditions. However, non-ideal systems also exist. The model can be revised to incorporate real-world complexities found in practical problems.

Sensitivity Analysis: The implementation of computational numerical methods enables the testing of the system with various inputs to assess its sensitivity.
Common tasks that are often solved using numerical methods include finding roots, solving linear equations, performing numerical interpolation, and applying calculus. Numerical methods are widely applied across various engineering disciplines, including kinematics, temperature distribution, and circuitry.

1.4 Modern Applications and Computer Software

The use of numerical analysis and mathematical modelling has become crucial in various aspects of contemporary society. Advanced numerical analysis software is now being integrated into widely-used software packages, such as spreadsheet programs. This enables individuals to engage in modelling tasks, even if they lack knowledge of the underlying mathematics. To address this challenge, it is necessary to develop numerical analysis software that is dependable, streamlined, and precise. Additionally, it is important to design problem-solving environments (PSE) that facilitate the modelling of specific scenarios with ease. The PSE for a given problem area typically relies on well-developed theoretical mathematical models, which are accessible to the user via a user-friendly graphical interface. The development of software tools has made significant progress in certain fields, such as computer aided design of structures. However, other areas are still facing challenges in creating precise mathematical models and the necessary tools for their solution, such as atmospheric modelling.

1.4.1 Some Application Areas

Computer aided design (CAD) and computer aided manufacturing (CAM) are significant fields in engineering, with the development of sophisticated PSEs. Mathematical models require a diverse range of numerical analysis for their solution. The models are derived from the fundamental Newtonian laws of mechanics. There exists a range of potential models, and ongoing research is dedicated to their development. One crucial subject in CAD is the modelling of dynamic motion in mechanical systems. The mathematical model consists of systems of ordinary differential equations and algebraic equations, typically nonlinear in nature. The analysis of these mixed systems, known as differential-algebraic systems, is challenging yet crucial for accurately modelling moving mechanical systems. Developing simulators for various vehicles involves solving real-time differential-algebraic systems.

Studying atmospheric modelling is crucial in order to accurately simulate the dynamics of the Earth's atmosphere and gain insights into the potential impact of human activities on our environment. Several variables must be introduced. The variables in question are the velocity, pressure, and temperature in the atmosphere at a given position and time. Furthermore, it is imperative to conduct a comprehensive analysis of the diverse array of chemicals present in the atmosphere and their intricate interplay. This includes but is not limited to ozone, various chemical pollutants, carbon dioxide, and other relevant compounds. The equations used to study $v(x, y, z, t), p(x, y, z, t)$, and $T(x, y, z, t)$ are partial differential equations. The chemical kinetic interactions of the various chemicals are described using challenging ordinary differential equations. Various numerical analysis procedures are employed in atmospheric modelling, such as computational fluid mechanics and solving differential equations. For the numerical solution of partial differential equations.

In the realm of modern business, optimisation methods play a crucial role in determining the most efficient allocation of resources. These problems encompass inventory control, scheduling, optimal location of manufacturing and storage facilities, investment strategies, and more.

1.4.2 Computer Software

The availability of software for implementing common numerical analysis procedures is crucial. If the information is intended for widespread use, it must be dependable, precise, and streamlined. In addition, it should be written in a manner that allows for easy transferability across various computer systems. Since the 1970s, several government-sponsored research initiatives have focused on developing top-notch numerical analysis software for specific problem domains. An example of a recent project in this field is the LAPACK project, which includes advanced programs for fundamental problems in numerical linear algebra. There are two significant online numerical analysis libraries that host numerous large-scale numerical analysis programming projects. These libraries can be accessed through the following internet sites:

Table 1.1 Internet sites

Name	URL	Location
Netlib	www.netlib.org	Oak Ridge National Laboratory
GAMS	gams.nist.gov	National Institute of Standards and Technology

Fortran remains the preferred programming language for implementing numerical analysis methods. It is regularly updated to adapt to evolving requirements, with Fortran 95 being the latest standard. There are several other significant programming languages, including C, C++, and Java. One alternative method of providing numerical analysis programs and programming tools is through the development of comprehensive problem-solving environments. These environments encompass numerical, programming, and graphical tools, and often include advanced numerical analysis tools capable of addressing a wide range of fundamental problems. One widely recognised software in this field is MATLAB (The Mathworks, Inc.), a commercial package that has gained significant popularity for its effectiveness in numerical computing. There are two widely used commercial packages for analytical mathematics computing: Maple (Waterloo Maple, Inc.) and Mathematica (Wolfram Research, Inc.).

CHAPTER 2

Mathematical Preliminaries

2.1 Review of Calculus

This chapter provides an overview of the important calculus results often used in this course. Only essential definitions and key theorems, along with a brief outline of their proofs, are included. It is expected that the readers have prior knowledge of a basic calculus course. Section 1 provides a clear definition of continuity for a function and presents a rigorous proof of the intermediate value theorem. This theorem is fundamental for determining an initial approximation in iterative techniques used to solve nonlinear equations in chapter 3. The section 2 of the material covers the derivative of a function, Rolle's theorem, and the mean-value theorem for derivatives. The mean-value theorem for integration is covered in section 4. These two theorems play a crucial role in deriving the truncation error for numerical methods. In section 5, Taylor's theorem is discussed, which is crucial for deriving and analysing the errors of various numerical methods covered in this course.

In this chapter, we will be using the notation $[a, b]$ to represent a closed interval and (a, b) to represent an open interval. These notations are commonly used in mathematics to denote a range of values. It is important to note that the values of a and b must be finite real numbers such that $a < b$.

2.2 Continuity of a Function

Definition 2.1 (Continuity).

A function $f: \mathbb{R} \to \mathbb{R}$ is said to be continuous at a point $x_0 \in \mathbb{R}$ if

$$\lim_{x \to x_0} f(x) = f(x_0) \tag{2.1}$$

In other words, for any given $\epsilon > 0$, there exists a $\delta > 0$ such that

$$|f(x) - f(x_0)| < \epsilon \text{ whenever } |x - x_0| < \delta\,. \tag{2.2}$$

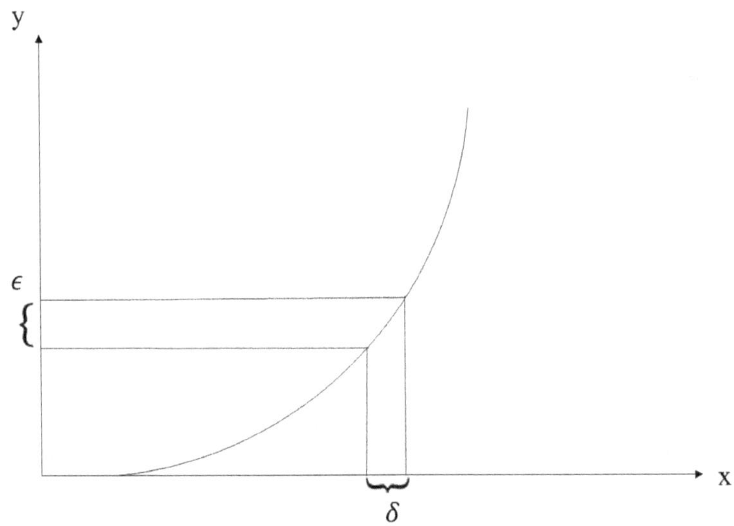

Figure 0.1 $y = x^2$

Example 2.1 Consider the function $f(x) = x^2$. Clearly, $f(x) = x^2 \to x_0^2$ when $x \to x_0$. Thus, this function is continuous. Let us now check the condition (2.2). We have,

$$\begin{aligned}|f(x) - f(x_0)| &= |x^2 - x_0^2| = |x + x_0||x - x_0| \\ &= |x - x_0 + 2x_0||x - x_0| \\ &\leq |x - x_0|(|x - x_0| + 2|x_0|)\,.\end{aligned}$$

For any given $\epsilon > 0$, choose $0 < \delta < -|x_0| + \sqrt{x_0^2 + \epsilon} > 0$ to get (2.2) as required. An illustration of this example is depicted in figure 2-1.

It is important to observe that the δ in the previous example is dependent upon x_0. For a continuous function f, if for any given $\epsilon > 0$, the corresponding δ is independent of x_0, then the function is classified as uniformly continuous.

Theorem 2.2 (Intermediate-Value Theorem)

Let $f(x)$ be a continuous function on the interval $[a, b]$. If $f(x_1) < \alpha < f(x_2)$ for some number α and some $x_1, x_2 \in [a, b]$ with $x_1 < x_2$, then

$f(\xi) = k$, for some $\xi \in [a, b]$.

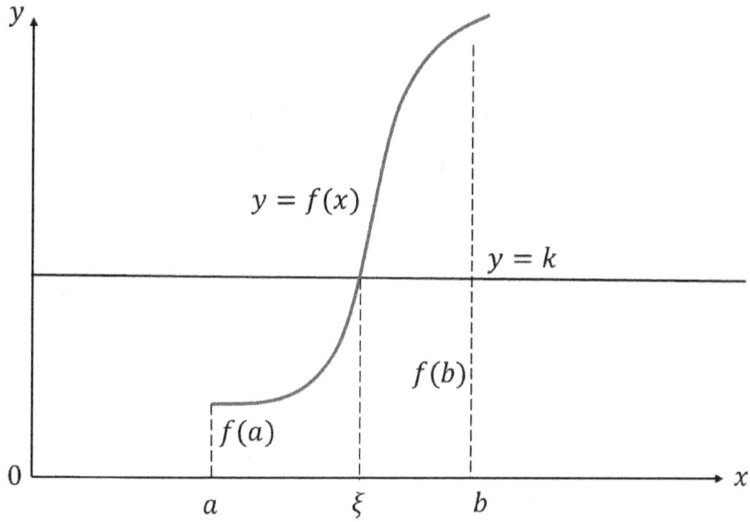

Figure 0.2 Intermediate Value Theorem

2.3 Differentiation of a Function

Definition 2.3 (Differentiation).

A function $f: (a, b) \to \mathbb{R}$ is said to be differentiable at a point $c \in (a, b)$ if the limit

$$\lim_{h \to 0} \frac{f(c + h) - f(c)}{h}$$

exists. In this case, the value of the limit is denoted by $f'(c)$ and is called the derivative of f at c. The function f is said to be differentiable in (a, b) if it is differentiable at every point in (a, b).

Remark 2.4. Two additional methods exist for defining the derivative of a continuous function $f: (a, b) \to \mathbb{R}$.

Here, we will outline the three equivalent definitions.

$$f'(c) = \lim_{h \to 0} \frac{f(c + h) - f(c)}{h},$$

$$f'(c) = \lim_{h \to 0} \frac{f(c) - f(c-h)}{h},$$
$$f'(c) = \lim_{h \to 0} \frac{f(c+h) - f(c-h)}{2h},$$

where $c \in (a, b)$. For any fixed $h > 0$, the formulae

$$D_h^+ f(c) := \frac{f(c+h) - f(c)}{h}, \tag{2.3}$$

$$D_h^- f(c) := \frac{f(c) - f(c-h)}{h}, \tag{2.4}$$

$$D_h^0 f(c) := \frac{f(c+h) - f(c-h)}{h} \tag{2.5}$$

The formulas used for this purpose are the forward difference, backward difference, and central difference formulae.

Theorem 2.4 (Rolle's Theorem).

Let $f(x)$ be continuous on the bounded interval $[a, b]$ and differentiable on (a, b). If $f(a) = f(b)$, then

$$f'(\xi) = 0, \qquad \text{for some } \xi \in (a, b).$$

Proof: Let $m, M \in [a, b]$ be such that

$f(m) = \min\{f(x) : x \in [a, b]\}$ and $f(M) = \max\{f(x) : x \in [a, b]\}$.

If either m or M is an interior point of $[a, b]$, then the result follows from the problem 11. Otherwise, both m and M are end points of $[a, b]$ and hence $f(m) = f(M)$. Thus, the maximum and the minimum values of f on $[a, b]$ coincide. Hence, f is constant on $[a, b]$, and therefore, $f'(x) = 0$ for every $x \in (a, b)$.

Theorem 2.5 (Mean-Value Theorem for Derivatives).

If $f(x)$ is continuous on a bounded interval $[a, b]$ (with $a \neq b$) and differentiable on (a, b), then

$$\frac{f(b)-f(a)}{b-a}=f'(\xi),$$
for some $\xi \in (a,b)$.

Proof: Let's assume that the function g is non-negative over the interval $[a,b]$. Next, we have

$$m\int_a^b g(x)dx \le \int_a^b f(x)g(x)dx \le M\int_a^b g(x)dx,$$

where m and M represent the minimum and maximum values of f within the interval $[a,b]$.

If $\int_a^b g(x)dx = 0$, then we have

$$\int_a^b f(x)g(x)dx = 0$$

The result in this case is straightforward. Let's assume the contrary and proceed by dividing both sides of the given inequality by $\int_a^b g(x)dx$ to get

$$m \le A(f) \le M,$$

Where

$$A(f)=\frac{1}{\int_a^b g(x)dx}\int_a^b f(x)g(x)dx.$$

Given the continuity of f, the intermediate value theorem implies the existence of a $\xi \in [a,b]$ such that $A(f)=f(\xi)$. This result serves as a proof for the theorem.

When g is non-positive, replace g by $-g$ and the same argument as above proves the theorem.

In the realm of calculus, the significance of this outcome may appear somewhat unclear, to say the least. From a numerical methods and analysis perspective, the Mean Value Theorem (MVT) is highly significant, ranking closely behind Taylor's Theorem,

$$f(x_1)-f(x_2)=f'(\xi)(x_1-x_2).$$

Hence, the Mean Value Theorem (MVT) enables us to substitute variations in argument values for variations in function values, provided we adjust for the function's derivative.

2.4 Integration of a Function

Theorem 2.6 (Mean-Value Theorem for Integrals).

Let $g(x)$ be a non-negative or non-positive integrable function on $[a, b]$. If $f(x)$ is continuous on $[a, b]$, then

$$\int_a^b f(x)g(x)dx = f(\xi) \int_a^b g(x)dx\,,$$
$$\text{for some } \xi \in (a, b)\,.$$

2.5 Taylor's Formula

If $f(x)$ has $n+1$ continuous derivatives on $[a, b]$ and c is some point in $[a, b]$, then for all $x \in [a, b]$

$$f(x) = f(c) + f'(c)(x-c) + \frac{f''(c)(x-c)^2}{2!} + \cdots + \frac{f^n(c)(x-c)^n}{n!} + R_{n+1}(x),$$

Where

$$R_{n+1}(x) = \frac{1}{n!}\int_c^x (x-t)^n f^{(n+1)}(t)dt\,.$$

Proof: The formula is proven using the method of induction.

(1) To begin, we will establish the formula for $n = 1$:

$$R_2(x) = f(x) - f(c) - f'(c)(x-c) = \int_c^x f'(t)dt - f'(c)\int_c^x dt$$
$$= \int_c^x \big(f'(t) - f'(c)\big)\,dt\,.$$

The final integral can be expressed as $\int_c^x udv$, with u representing $f'(t) - f(c)$, and $v = t - x$. Given the equations $du/dt = f''(t)$ and $dv/dt = 1$, we can apply the integration by parts method to obtain the following result:

$$R_2(x) = \int_c^x u dv = uv|_c^x - \int_c^x (t-x)f''(t)\,dt = \int_c^x (x-t)f''(t)\,dt$$

Since $u = 0$ when $t = c$ and $v = 0$ when $t = x$. This concludes the proof when $n = 1$.

(2) Let's assume that the formula is true for some value of n and now we will prove it for $n + 1$.The formula for $n + 1$ in Taylor's series can be expressed as:

$$R_{n+1}(x) = f(x) - \left(f(c) + f'(c)(x-c) + \frac{f''(c)(x-c)^2}{2!} \cdots + \frac{f^{n-1}(c)(x-c)^{n-1}}{(n-1)!} + \frac{f^n(c)(x-c)^n}{n!}\right)$$

$$= R_n(x) - \frac{f^n(c)(x-c)^n}{n!}$$

Given that the Taylor's formula holds for n, we can use the provided remainder formula for $R_n(x)$. Using the concept of identity

$$\frac{(x-c)^n}{n} = \int_c^x (x-t)^{n-1} dt \ ,$$

We get,

$$R_{n+1}(x) = \frac{1}{(n-1)!}\int_c^x (x-t)^{(n-1)} f^n(t) dt - \frac{f^n(c)}{(n-1)!}\int_c^x (x-t)^{n-1} dt = \frac{1}{(n-1)!}\int_c^x (x-t)^{(n-1)}[f^n(t) - f^n(c)]dt\,.$$

The final integral can be expressed as $\int_c^x u dv$, where $u = f^n(t) - f^n(c)$ and

$$v = -{}^{(x-t)^n}/_{n}.$$

It is worth mentioning that when $t = c, u = 0$,
and when $t = x, v = 0$. We obtain the result through the application of the integration by parts method.

$$R_{n+1}(x) = \frac{1}{(n-1)!}\int_c^x u dv = -\frac{1}{(n-1)!}\int_c^x u dv$$
$$= \frac{1}{n!}\int_c^x (x-t)^n f^{(n+1)}(t) dt .$$

The inductive step from n to $n+1$ has been successfully demonstrated, thus confirming the validity of the theorem for all $n \geq 1$.

Remark 2.7. It is important to note that as x approaches c, the remainder $R_{n+1}(x)$ tends to zero. Therefore, the Taylor's formula (without remainder) can be utilised to obtain an approximate value of f at any given point x within a close vicinity of c, provided that the values of f and all its n derivatives are known at c.

Here, we present the two-dimensional Taylor's formula and defer the proof for further study. In order to provide a clear and straightforward explanation, we will present the formula for $n = 1$, with the understanding that it can be easily extended to other cases.

Using Taylor's theorem, we obtain the following standard formulas:

$$e^x = 1 + x + \frac{x^2}{2} + \cdots + \frac{x^n}{n!} + \frac{x^{n+1}}{(n+1)!} e^{\xi_x} \tag{2.6}$$

$$\cos(x) = 1 - \frac{x^2}{2!} + \frac{x^4}{4!} - \cdots + (-1)^n \frac{x^{2n}}{(2n)!} + (-1)^{n+1} \frac{x^{2n+2}}{(2n+2)!} \cos(\xi_x) \tag{2.7}$$

$$\sin(x) = x - \frac{x^3}{3!} + \frac{x^5}{5!} - \cdots + (-1)^{n-1} \frac{x^{2n-1}}{(2n-1)!} + (-1)^n \frac{x^{2n+1}}{(2n+1)!} \cos(\xi_x) \tag{2.8}$$

$$(1+x)^\alpha = 1 + \binom{\alpha}{1} x + \binom{\alpha}{2} x^2 + \cdots + \binom{\alpha}{n} x^n + \binom{\alpha}{n+1} \frac{x^{n+1}}{(1+\xi_x)^{n+1-\alpha}} \tag{2.9}$$

With

$$\binom{\alpha}{k} = \frac{\alpha(\alpha-1)\cdots(\alpha-k+1)}{k!}, \qquad k = 1,2,3,\cdots$$

for any real number α . For all cases, the unknown point ξ_x is located between x and 0.

An important special case of (2.9) is

$$\frac{1}{1-x} = 1 + x + x^2 + \cdots + x^n + \frac{x^{n+1}}{1-x}, \qquad x \neq 1 \tag{2.10}$$

Example 2.2 Let $f(x) = e^{-x^2}$

Let's replace x by $-x^2$ in (2.6) to obtain

$$e^{-x^2} = 1 - x^2 + \frac{x^2}{2!} + \cdots + (-1)^n \frac{x^{2n}}{n!} + (-1)^{n+1} \frac{x^{2n+2}}{(n+1)!} e^{\xi_x}$$

With $-x^2 \leq \xi_x \leq 0$.

Example 2.3 Let $f(x) = \tan^{-1}(x)$.

Now, begin by establishing $x = -u^2$ in equation (2.10)

$$\frac{1}{1+u^2} = 1 - u^2 + u^4 - \cdots + (-1)^n u^{2n} + (-1)^{n+1} \frac{u^{2n+2}}{1+u^2}$$

Integrate over $[0, x]$ to get

$$\tan^{-1}(x) = x - \frac{x^3}{3} + \frac{x^5}{5} - \cdots + (-1)^n \frac{x^{2n+1}}{2n+1} + (-1)^{n+1} \int_0^x \frac{u^{2n+2}}{1+u^2} du$$

Applying the integral mean value theorem

$$\int_0^x \frac{u^{2n+2}}{1+u^2} du = \frac{x^{2n+3}}{2n+3} \cdot \frac{1}{1+\xi_x^{\,2}}$$

With ξ_x between 0 and x.

Example 2.4 Let $f(x) = \int_0^1 \sin(xt)\, dt$.
Using (2.8)

$$f(x) = \int_0^1 \left[xt - \frac{x^3t^3}{3!} + \cdots + (-1)^{n-1}\frac{(xt)^{2n-1}}{(2n-1)!} + (-1)^n \frac{(xt)^{2n+1}}{(2n+1)!}\cos(\xi_{xt}) \right] dt$$

$$= \sum_{j=1}^{n} (-1)^{j-1}\frac{x^{2j-1}}{(2j)!} + (-1)^n \frac{x^{2n+1}}{(2n+1)!}\int_0^1 t^{2n+1}\cos(\xi_{xt})\, dt$$

With ξ_{xt} between 0 and xt. The integral in the remainder is easily bounded by $1/(2n + 2)$; but we can also convert it to a simpler form. Although it wasn't proved, it can be shown that $\cos(\xi_{xt})$ is a continuous function of t. Then applying the integral mean value theorem

$$\int_0^1 \sin(xt)\, dt = \sum_{j=1}^{n} (-1)^{j-1}\frac{x^{2j-1}}{(2j)!} + (-1)^n \frac{x^{2n+1}}{(2n+1)!}\cos(\xi_{xt})$$

for some ξ_x between 0 and x.

Theorem 2.8 (Taylor's Formula in 2-Dimensions).

In the region D of the point (a, b), if $f(x, y)$ is a function of the two independent variables x and y that is continuous and has continuous first and second partial derivatives, then

$$f(x,y) = f(a,b) + f_x(a,b)(x-a) + f_y(a,b)(y-b) + R_2(x,y),$$

for all $(x, y) \in D$, where,

$$R_2(x,y) = \frac{f_{xx}(\xi,\eta)(x-a)^2}{2} + f_{xy}(\xi,\eta)(x-a)(y-b) + \frac{f_{xy}(\xi,\eta)(y-b)^2}{2},$$

for some $(\xi, \eta) \in D$ depending on (x, y) and the subscripts of f denote partial differentiation.

Lemma 2.9 (Chain Rule).

If the function $f(x_1, x_2, \dots, x_n)$ possesses continuous first partial derivatives in relation to each of its variables, and $x_1 = x_1(t), x_2 = x_2(t), \dots, x_n = x_n(t)$ are continuously differentiable functions of t, then $g(t) = f(x_1(t), x_2(t), \ldots, x_n(t))$ is likewise continuously differentiable, and

$$g'(t) = \frac{\partial f}{\partial x_1}x'_1(t) + \frac{\partial f}{\partial x_2}x'_2(t) + \cdots + \frac{\partial f}{\partial x_n}x'_n(t)\,.$$

Notes

This chapter incorporates material from Ghorpade and Limaye (2006) and Apostal Volume 1 (2002), including several exercise problems. These books may be consulted for further details on the calculus applied throughout this course.

2.6 Exercises

1. Explain why each of the following functions is continuous or discontinuous.
 (a) The temperature at a specific location as a function of time.
 (b) The temperature at a specific time as a function of the distance from a fixed point.

2. Check the continuity of the function f in each of the following cases:

(a) $f(x) = \begin{cases} x^2 & \text{if } x < 1 \\ \sqrt{x} & \text{if } x \geq 1 \end{cases}$,

(b) $f(x) = \begin{cases} -x & \text{if } x < 1 \\ x & \text{if } x \geq 1 \end{cases}$,

(c) $f(x) = \begin{cases} 0 & \text{if } x \text{ is rational} \\ 1 & \text{if } x \text{ is rational} \end{cases}$,

3. Show that the equation

$$\sin x + x^2 = 1$$

has at least one solution in the interval $[0,1]$.

4. Let $c \in (a, b)$ and $f: (a, b) \to R$ is differentiable at c. If c is a local extremum (maximum or minimum) of f, then show that $f'(c) = 0$.

5. Let $f(x) = 1 - x^{2/3}$. Show that $f(1) = f(-1) = 0$, but that $f'(x)$ is never zero in the interval $[-1,1]$. Explain how this is possible, in view of Rolle's theorem.

6. Show that the function $f(x) = \cos x$ for all $x \in R$ is continuous by choosing an appropriate $\delta > 0$ for a given $\epsilon > 0$ as in the definition 2.1.

7. For $f(x) = x^2$, find the point ξ specified by the mean-value theorem for derivatives. Verify that this point lies in the interval (a, b).

8. In the mean-value theorem for integrals, let $f(x) = e^x$, $g(x) = x$, $[a, b] = [0,1]$. Find the point ξ specified by the theorem and verify that this point lies in the interval $(0,1)$.

9. Find the Taylor's expansion for $f(x) = \sqrt{x+1}$ upto $n = 2$ (ie. the Taylor's polynomial of order 2) with remainder $R_3(x)$ about $c = 0$.

10. Show that the remainder $R_{n+1}(x)$ in the Taylor's expansion of a $n+1$ continuously differentiable function f can be written as

$$R_{n+1}(x) = \frac{(x-c)^{n+1}}{(n+1)!} f^{(n+1)}(\xi),$$

Where $\xi \in (c, x)$.

CHAPTER 3

Error Sources, Propagation and Analysis in Numerical Methods

3.1 Definitions of Errors

Numerical errors occur due to the application of approximations in representing precise mathematical operations and quantities. Truncation errors arise from the use of approximations to represent precise mathematical procedures, while round-off errors occur when numbers with limited significant figures are employed to represent exact values. The relationship between the exact result and the approximation for both types can be expressed as

$$True\ value = approximation + error \tag{3.1}$$

Rearranging Eq. (3.1) reveals that the numerical error corresponds to the difference between the true value and the approximation,

$$E_t = true\ value - approximation \tag{3.2}$$

In this context, E_t denotes the exact value of the error. The subscript t indicates that this is the "true" error. This contrasts with other cases, where a "approximate" estimate of the error is required.

This definition fails to consider the order of magnitude of the value being analyzed. An error of one centimeter holds greater significance in the context of measuring a rivet compared to measuring a bridge. One method to assess the magnitudes of the evaluated quantities is to normalize the error relative to the true value, as demonstrated in

$$\text{True fractional relative error} = \frac{\text{true error}}{\text{true value}}$$

According to Eq. (3.2), $error = true\ value - approximation$. The relative error may also be multiplied by 100 percent for expression as

$$\varepsilon_t = \frac{\text{true error}}{\text{true value}} 100\% \tag{3.3}$$

Where, ε_t designates the true percent relative error.

Example 3.1 Suppose that you have the task of measuring the lengths of a bridge and a rivet and come up with 9999 and 9 cm, respectively. If the true values are 10,000 and 10 cm, respectively, compute (a) the true error and (b) the true percent relative error for each case.

Solution

(a) The error for measuring the bridge is [Eq. (3.2)]

$$E_t = 10{,}000 - 9999 = 1 \text{ cm}$$

and for the rivet it is

$$E_t = 10 - 9 = 1 \text{ cm}$$

(b) The percent relative error for the bridge is [Eq. (3.3)]

$$\varepsilon_t = \frac{1}{10{,}000} 100\% = 0.01\%$$

and for the rivet it is

$$\varepsilon_t = \frac{1}{10} 100\% = 10\%$$

Therefore, despite both measurements exhibiting an error of 1 cm, the relative error associated with the rivet is significantly larger. In conclusion, our measurement of the bridge is satisfactory, while our estimate for the rivet is lacking.

Truncation error refers to the error that arises from approximating a smooth function by limiting its Taylor series representation to a finite number of terms.

Example 3.2 A second degree polynomial approximation to

$$f(x) = \sqrt{x+1}, \quad x \in [0,1]$$

using the Taylor series expansion about $x = 0$ is given by

$$f(x) \approx 1 + \frac{x}{2} - \frac{x^2}{8} + \frac{x^3}{16\left(\sqrt{1+\xi}\right)^5}$$

Therefore, the truncation error is given by $x^3/16\left(\sqrt{1+\xi}\right)^5$.

Remark 3.1. Let x_A be the approximation of the real number x. Then

$$E(x_A) := Error(x_A) = x - x_A \,. \tag{3.4}$$
$$E_a(x_A) := Absolute\ Error(x_A) = |E(x_A)| \tag{3.5}$$
$$E_r(x_A) := Relative\ Error(x_A) = {}^{E(x_A)}/_{x} \tag{3.6}$$

Example 3.3 If we denote the relative error in $fl(x)$ as $\epsilon > 0$, then we have

$$fl(x) = (1 - \epsilon)x \,, \tag{3.7}$$

where x is a real number.

3.2 Round-off Errors and Computer Arithmetic

Arithmetic executed by a calculator or computer differs from the arithmetic encountered in algebra and calculus courses. One would anticipate the presence of arithmetic equations such as $2 + 2 = 4$, $4 \cdot 8 = 32$, and $\left(\sqrt{3}\right)^2 = 3$. Nevertheless, using computer arithmetic, we anticipate obtaining exact outcomes for $2 + 2 = 4$ and $4 \cdot 8 = 32$, but we will not have precisely $\left(\sqrt{3}\right)^2 = 3$. In order to comprehend the validity of this statement, it is necessary to delve into the realm of finite-digit addition.

In traditional mathematics, we allow for numbers that possess an infinite number of digits. The arithmetic utilized in this context defines $\sqrt{3}$ as the unique positive number that, when squared, yields the integer 3. In computational contexts, each representable number is constrained to a fixed and finite number of digits. This indicates that only certain rational numbers can be represented with precision. since $\sqrt{3}$ is irrational, it is represented approximately, resulting in a value whose square will not equal exactly 3, though it will typically be sufficiently close to 3 for most practical purposes. In many instances, machine arithmetic is adequate and goes unnoticed; however, discrepancies can occasionally lead to issues.

Round-off error refers to the discrepancy that arises when a calculator or computer executes calculations involving real numbers. The phenomenon arises due to the finite digit representation of numbers in machine arithmetic, leading to calculations based on approximate representations of the actual values. In computing, a limited subset of the real number system is employed to represent all real numbers. This subset comprises solely rational numbers, encompassing both positive and negative values, and retains the fractional component along with an exponential component.

3.3 Floating-Point Form of Numbers

Real numbers are represented in floating-point form on a computer, which is to be introduced in this section.

Definition 3.2 (Floating-Point Form).

Let x represent a non-zero real number. A floating-point number with n digits in base β is represented as

$$fl(x) = (-1)^s \times (.d_1 d_2 \cdots d_n)_\beta \times \beta^e \tag{3.8}$$

Where

$$(.d_1 d_2 \cdots d_n)_\beta = \frac{d_1}{\beta} + \frac{d_2}{\beta^2} + \cdots + \frac{d_n}{\beta^n} \tag{3.9}$$

is a β −fraction known as the **mantissa** or **significand**, $s = 1$ or 0 is called the **sign** and e is an integer called the **exponent**. The number β is referred to as the **radix**, while the point preceding d_1 in (3.8) is termed the radix point.

Remark 3.3. When $\beta = 2$, the representation (3.8) is known as **binary** floating-point, whereas when $\beta = 10$, it is referred to as **decimal** floating-point representation.

It is important to note that the floating-point representation includes a finite number of digits (3.8), whereas a real number might have an infinite sequence of digits; for example, $1/3 = 0.33333......$ Consequently, representation (3.8) serves only as an approximation of a real number.

Definition 3.4 (Normalization).

A floating-point number is considered **normalized** if either $d_1 \neq 0$ or $d_1 = d_2 = \cdots = d_n = 0$.

Example 3.4 Examples of real numbers in decimal floating point representation are provided below.

1. The real number $x = 6.238$ can be expressed as $6.238 = (-1)^0 \times 0.6238 \times 10^1$. In this representation, the parameters are defined as $s = 0, \beta = 10, e = 1, d_1 = 6, d_2 = 2, d_3 = 3$, and $d_4 = 8$. This representation is the normalized floating-point representation.

2. The real number $x = -0.0014$ can be expressed in decimal floating-point representation as $-0.0014 = (-1)^1 \times 0.0014 \times 10^1$, which does not conform to the normalised form. This representation is not in normalized form. The normalized representation is expressed as $x = (-1)^1 \times 0.14 \times 10^{-2}$.

Definition 3.5 (Overflow and Underflow).

The exponent e is constrained within the interval

$$m < e < M \tag{3.10}$$

In the calculation process, if a computed number has an exponent $e > M$, it is classified as memory **overflow**; conversely, if $e < m$, it is termed memory **underflow**.

Remark 3.6. In cases of overflow, a computer typically generates nonsensical results or outputs the symbol NaN, indicating that the resultant quantity from the calculation is 'not a number'. The symbol ∞ is alternatively represented as NaN on certain computer systems. The underflow is less critical in this scenario, as the computer will interpret the number as zero.

Remark 3.7. The floating-point representation (3.8) of a number is subject to two constraints: the number of digits n in the mantissa and the range of e. The variable n is referred to as the **precision** or **length** of the floating-point representation.

Example 3.5 The IEEE 754 standard for floating-point arithmetic is the predominant standard for floating-point computation, adopted by numerous hardware implementations, including CPUs and FPUs, as well as various software applications. Numerous programming languages permit or mandate the execution of certain or all arithmetic operations utilizing IEEE 754 formats and procedures. The IEEE 754 floating-point representation of a binary number x is defined as

$$fl(x) = (-1)^s \times (1.a_1a_2 \cdots a_n)_2 \times 2^e\,, \tag{3.11}$$

Where $a_1, \cdots, a_n$ are either 1 or 0. The IEEE 754 standard always uses binary operations.

The difference between the representation provided in (3.8) and the one presented here should be noted. In binary representation, the digit preceding the binary point is consistently 1; consequently, this information does not need to be stored in computer memory. This form of representation was preferred over (3.8) for this reason.

The IEEE single precision floating-point format takes 4 bytes (32 bits) for number representation. Among the 32 bits, 24 bits are designated for the mantissa (with each binary digit requiring 1 bit), 1 bit is allocated for the sign (s), and the remaining 8 bits are reserved for the exponent. The storage scheme is presented as follows:

$$|(sign)\ b_1|\ (exponent)\ b_2b_3 \cdots b_9\ |\ (mantissa)\ b_{10}b_{11} \cdots b_{32}|$$

It is important to note that only 23 bits are allocated for the mantissa. The digit 1 preceding the binary point in (3.11) is not retained in memory and is instead incorporated during the calculation process. Rather than using the exponent e, we represent the non-negative integer $E = (b_2b_3 \cdots b_9)_2$ and establish $e = E - 127$. If all b_i's $(i = 2{,}9)$ are zero, then $E = (0)_{10}$; conversely, if all b_i's are one, then $E = (255)_{10}$. Furthermore, one space corresponding to $e = 128$ (or $E = 255$) is allocated for ∞ or NaN, contingent upon whether $b_{10} = \cdots = b_{32} = 0$ or otherwise. In IEEE 754, we have, $-126 \le e \le 127$. It is important to note that the range does not include -127, as this value is designated for numbers that are not represented in other ways (refer to Atkinson and Han, 2004, for further details). Additionally, there is one memory space allocated for NaN. The decimal number zero requires a specific representation, denoted as $E = 0$ (i.e., $b_2 = b_9 = 0$), with $b_1 = 0$ and $b_{10} = \cdots = b_{32} = 0$.

In representation (3.11), the value of s is contained in b_1, while the positive integer $E = e + 127$ is stored in bits b_2 through b_9. The sequence of digits a_1, a_2, and a_{23} is stored in bits b_{10} to b_{32}. The leading binary digit 1 in the mantissa is not retained in memory. However, this information is incorporated into the mantissa when a floating-point number x is retrieved from memory and subjected to an arithmetic operation. In the IEEE single precision storage system, overflow occurs for real numbers $|x| > x_{max}$, where

$$x_{max} = 1.11 \cdots 1 \times 2^{127} \approx 3.40 \times 10^{38} .$$

The IEEE double precision floating-point representation allows for a precision of 53 binary digits and the exponent e is limited by $-1023 \leq e \leq 1023$.

Example 3.6 Consider evaluating $x = s^{10}$ on an IBM mainframe computer. Then there will be an underflow error if

$$s^{10} < 16^{-65}$$

Thus $x = s^{10}$ will be set to zero if $|s| < 1.49 \times 10^{-8}$. Also, there will be an overflow error if

$$s^{10} > 16^{-65}$$

or equivalently, $|s| < 3.9 \times 10^7$

3.4 Chopping and Rounding a Number

Any real number x can be expressed as

$$x = (-1)^s \times (.d_1 d_2 \cdots d_n d_{n+1} \cdots)_\beta \times \beta^e , \qquad (3.12)$$

where $d_1 \neq 0$ or $d_2 = d_3 = \cdots = 0$, $s = 0$ or 1, and e satisfies the conditions outlined in (2.3).

The floating-point representation given in (3.10) serves as an approximation of this exact representation. We denote this approximation of x as $fl(x)$. Two methods exist for producing $fl(x)$ from x, as defined below.

Definition 3.8 (Chopped and Rounded Numbers).

The chopped machine approximation of x is given by

$$fl(x) = (-1)^s \times (.d_1 d_2 \cdots d_n)_\beta \times \beta^e , \qquad (3.13)$$

The rounded machine approximation of x is given by

$$fl(x) = \begin{cases} (-1)^s \times (.d_1 d_2 \cdots d_n)_\beta \times \beta^e & , 0 \leq d_{n+1} < \frac{\beta}{2} \\ (-1)^s \times (.d_1 d_2 \cdots d_n + 1)_\beta \times \beta^e & , \frac{\beta}{2} \leq d_{n+1} < \beta \end{cases} \tag{3.14}$$

Example 3.7 Determine the five-digit (a) chopping and (b) rounding values of the irrational number π.

Solution. The number π has an infinite decimal expansion of the form $\pi = 3.14159265 \cdots$

Written in normalized decimal form, we have

$$\pi = 3.14159265 \cdots \times 10^1$$

(a) The floating-point form of π using five-digit chopping is

$$fl(\pi) = 3.1415 \times 10^1$$

(b) The sixth digit of the decimal expansion of π is a 9, so the floating-point form of π using five-digit rounding is

$$fl(\pi) = (0.31415 + 0.00001) \times 10^1 = 3.1416$$

Example 3.8 For PRIME computers in single precision,

$$M = 2^{23} = 8388608 .$$

Therefore, all integers with six decimal digits and the majority with seven decimal digits can be accurately represented in the single precision floating-point format. For the unit round,

$$\delta = 2^{-22} = 2.38 \times 10^{-7} \qquad \text{chopped arithmetic}$$
$$\delta = 2^{-23} = 1.19 \times 10^{-7} \qquad \text{rounded arithmetic}$$

Users of the PRIME have access to both chopped and rounded arithmetic for single precision calculations. In nearly all cases, rounded arithmetic is significantly more beneficial than chopped arithmetic.

3.5 Propagation of Error

An error, once committed, influences subsequent results as it propagates through later calculations. Initially, we examine the impact of utilizing approximate numbers in place of actual numbers, followed by an analysis of function evaluation.

Let x_A and y_A represent the values utilized in the computation, while x_T and y_T denote the corresponding true values. We will now examine the propagation of error through the four fundamental arithmetic operations.

Propagated error in addition and subtraction

Let $x_T = x_A + \epsilon$ and $y_T = y_A + \eta$, where both are positive numbers. The relative error $E_r(x_A \pm y_A)$ is expressed as

$$
\begin{aligned}
E_r(x_A \pm y_A) &= \frac{(x_T \pm y_T) - (x_A \pm y_A)}{(x_T \pm y_T)} \\
&= \frac{(x_T \pm y_T) - (x_T - \epsilon \pm (y_T - \eta))}{(x_T \pm y_T)} = \frac{\epsilon \pm \eta}{(x_T \pm y_T)}.
\end{aligned}
$$

This shows that relative error propagate slowly with addition, whereas amplifies drastically with subtraction when $x_T \approx y_T$.

Propagated error in multiplication

The relative error $E_r(x_A \times y_A)$ is expressed as

$$
\begin{aligned}
E_r(x_A \times y_A) &= \frac{(x_T \times y_T) - (x_A \times y_A)}{(x_T \times y_T)} \\
&= \frac{(x_T \times y_T) - ((x_T - \epsilon) \times (y_T - \eta))}{(x_T \times y_T)} \\
&= \frac{\eta x_T + \epsilon y_T - \epsilon\eta}{(x_T \times y_T)} = \frac{\epsilon}{x_T} + \frac{\eta}{y_T} - \left(\frac{\epsilon}{x_T}\right)\left(\frac{\eta}{y_T}\right) \\
&= E_r(x_A) + E_r(y_A) - E_r(x_A) + E_r(y_A).
\end{aligned}
$$

This indicates that relative error propagate slowly with multiplication.

Propagated error in division

The relative error $E_r(x_A/y_A)$ is expressed as

$$
\begin{aligned}
E_r(x_A/y_A) &= \frac{(x_T/y_T) - (x_A/y_A)}{(x_T/y_T)} = \frac{(x_T/y_T) - ((x_T - \epsilon)/(y_T - \eta))}{(x_T/y_T)} \\
&= \frac{x_T(y_T - \eta) - y_T(x_T - \epsilon)}{x_T(y_T - \eta)} = \frac{y_T\epsilon - x_T\eta}{\square} \\
&= \frac{y_T}{y_T - \eta}\left(E_r(x_A) - E_r(y_A)\right)
\end{aligned}
$$

This indicates that relative error propagates slowly during division, except when Er(YA) is approximately equal to 1. However, this situation is highly unlikely as we anticipate the error to be minimal, approaching zero, in which case the right-hand side is approximately equal to $E_r(x_A) - E_r(y_A)$.

Total calculation error

In floating-point arithmetic on a computer, the computation of $x_A \omega y_A$ (where ω represents one of the fundamental arithmetic operations: '$+$','$-$','$\times$', or '/') introduces an additional rounding or chopping error. The calculated value of $x_A\ \widehat{\omega} y_A$ will encompass the propagated error in addition to a rounding or chopping error. Let $\widehat{\omega}$ represent the complete operation executed on the computer, encompassing all rounding or chopping processes. The **total error** is expressed as

$$
\begin{aligned}
&(x_T \omega y_T) - (x_A\ \widehat{\omega} y_A) \\
&\qquad = [(x_T \omega y_T) - (x_A \omega y_A)] + [(x_A \omega y_A) - (x_A \widehat{\omega} y_A)]
\end{aligned}
$$

The initial term on the right represents the propagated error, whereas the subsequent term denotes the error resulting from rounding or chopping the calculated value $x_A \omega y_A$.

Propagated error in function evaluation

Let's evaluate $f(x)$ at the approximate value x_A rather than at x. Then consider the degree to which $f(x_A)$ approximates $f(x)$. Applying the mean-value theorem, we get

$$
f(x) - f(x_A) = f'(\xi)(x - x_A),
$$

where ξ represents an unknown point located between x and x_A. The relative error of $f(x)$ in relation to $f(x_A)$ is expressed as

$$E_r(f(x)) = \frac{f'(\xi)}{f(x)}(x - x_A) = \frac{f'(\xi)}{f(x)} E_r(x). \tag{3.15}$$

Given that x_A and x are considered to be quite similar and ξ is positioned between x and x_A, we can proceed with the approximation.

$$f(x) - f(x_A) \approx f'(x)(x - x_A) \approx f'(x_A)(x - x_A).$$

Definition 3.9 (Condition number of a function).

The condition number of a function f at a point $x = c$ can be expressed as

$$\left|\frac{f'(c)}{f(c)} c\right| \tag{3.16}$$

Example 3.9 Consider the function

$$f(x) = \sqrt{x}, \text{ for all } x \in [0, \infty). \text{ Then}$$
$$f'(x) = {}^{1}/_{2\sqrt{x}}, \qquad \text{for all } x \in [0, \infty).$$

The condition number of f is

$$\left|\frac{f'(x)}{f(x)} x\right| = \frac{1}{2},$$
$$\text{for all } x \in [0, \infty).$$

From (3.15) we see that taking square roots is a **well-conditioned** process since it actually reduces the relative error.

Example 3.10 Consider the function

$$f(x) = \frac{10}{1 - x^2},$$
$$\text{for all } x \in \mathbb{R}$$

Then $f'(x) = 20x/(1 - x^2)^2$, so that

$$\left|\frac{f'(x)}{f(x)} x\right| = \left|\frac{(20x/(1 - x^2)^2)x}{10/(1 - x^2)}\right| = \frac{2x^2}{|1 - x^2|}$$

and this number can be quite large for $|x|$ near 1. Thus, for x near 1 or -1, this function is **ill-conditioned**, as it magnifies the relative error.

Definition 3.10 (Stability and Instability in Evaluating a Function).

Let n represent the number of steps required to evaluate the function $f(x)$. The overall evaluation of this function is considered unstable if at least one step is ill-conditioned. If all steps are well-conditioned, the process is considered stable.

Example 3.11 Consider the function

$$f(x) = \sqrt{x+1} - \sqrt{x}\,, \text{ for all } x \in [0, \infty).$$

For a sufficiently large x, the condition number of this function is

$$\left|\frac{f'(x)}{f(x)}x\right| = \frac{1}{2}\left|\frac{\frac{1}{\sqrt{x+1}} - \frac{1}{\sqrt{x}x}}{\sqrt{x+1} - \sqrt{x}}\right| = \frac{1}{2}\frac{x}{\sqrt{x+1}\sqrt{x}} \approx \frac{1}{2},$$

which is quite good. But, if we calculate $f(12345)$ in six digit rounding arithmetic, we find

$$f(12345) = \sqrt{12346} - \sqrt{12345} = 111.113 - 111.108 = 0.005$$

while, $f(12345) = 0.0045000326262775$1. The computed result exhibits a 10% error margin.

Let us examine the computational process. The process comprises four computational steps:

$$x_0 := 12345, \qquad x_1 := x_0 + 1, \qquad x_2 := \sqrt{x_1}\,, x_3 := \sqrt{x_0}\,,$$
$$x_4 := x_2 - x_3\,.$$

Now consider the last two step where we already computed x_2 and now going to compute x_3 and finally evaluate the function

$$f_3(t) = x_2 - t\,.$$

At this stage, the condition number for f_3 is given by

$$\left|\frac{f'(t)}{f(t)}t\right| = \left|\frac{t}{x_2 - t}\right|.$$

Consequently, f exhibits ill-conditioning as t approaches x_2. For $t \approx 111.11$, when $x_2 - t \approx 0.005$, the condition number for f_3 is approximately 22.222, which is over 40,000 times greater than the condition number of f itself. Consequently, the above process for evaluating the function $f(x)$ exhibits instability.

Let us rewrite the same function $f(x)$ as

$$f(x) = \frac{1}{\sqrt{x+1} + \sqrt{x}}.$$

In six digit rounding arithmetic, this gives

$$f(12345) = \frac{1}{\sqrt{12346} + \sqrt{12345}} = \frac{1}{222.221} = 0.0045002$$

which is in error by only 0.0003%. The computational process is

$$x_0 := 12345, \qquad x_1 := x_0 + 1, \qquad x_2 := \sqrt{x_1}\ , x_3 := \sqrt{x_0}\ ,$$
$$x_4 := x_2 + x_3\ , \qquad x_5 := \frac{1}{x_4}.$$

The condition number for each of the aforementioned steps is readily verifiable as being well-conditioned. For example, the final step establishes $f_3(t) = 1/(x_2 + t)$, and the condition number of this function is approximately

$$\left|\frac{f'(x)}{f(x)}x\right| = \left|\frac{t}{x_2 - t}\right| \approx \frac{1}{2}$$

for t sufficiently close to x_2. Therefore, this process of evaluating $f(x)$ is stable.

3.6 Exercises

1. Write the storage scheme for the IEEE double precision floating-point representation of a real number with the precision of 53 binary digits. Find the overflow limit (in binary numbers) in this case.

2. In a binary representation, if 2 bytes (ie., $2 \times 8 = 16$ bits) are used to represent a floating-point number with 8 bits used for the exponent. Then, as of IEEE 754 storage format, find the largest binary number that can be represented.

3. Let $p = 0.54617$ and $q = 0.54601$. Use four-digit arithmetic to approximate $p - q$ and determine the absolute and relative errors using (a) rounding and (b) chopping.

4. Find the largest interval in which p^* must lie to approximate p with relative error at most 10^{-4} for each value of p.

a) π b) e

c) $\sqrt{2}$ d) $\sqrt[3]{7}$

5. If $fl(x)$ is the machine approximated number of a real number x and ϵ is the corresponding relative error, then show that $fl(x) = (1 - \epsilon)x$.

6. Let x, y and z are the given machine approximated numbers. Show that the relative error in computing $x(y + z)$ is $\epsilon_1 + \epsilon_2 - \epsilon_1\epsilon_2$, where $\epsilon_1 = E_r(fl(y + z))$ and $\epsilon_2 = E_r\left(fl\left(xfl(y + z)\right)\right)$.

7. If the relative error of $fl(x)$ is ϵ, then show that

$$|\epsilon| \leq \beta^{-n+1} \quad (\text{for rounded } fl(x)),$$

where β is the radix and n is the number of digits in the machine approximated number.

8. Let x_A and y_A, the approximation to x and y, respectively, be such that the relative errors $E_r(x)$ and $E_r(y)$ are very much smaller than 1. Then show that

(i) $E_r(xy) \approx E_r(x) + E_r(y)$ and

(ii) $E_r(x/y) \approx E_r(x) - E_r(y)$. (This shows that relative errors propagate slowly with multiplication and division).

9. Find the condition number for the following functions (a) $f(x) = x^2$, (b) $f(x) = \pi^x$, (c) $f(x) = b^x$.

10. Given a value of $x_A = 2.5$ with an error of 0.01. Estimate the resulting error in the function $f(x) = x^3$.

11. Compute and interpret (find whether the functions are well or ill-conditioned) the condition number for (i) $f(x) = \tan x$, at $x = \frac{\pi}{2} + 0.1\left(\frac{\pi}{2}\right)$. (ii) $f(x) = \tan x$, at $x = \frac{\pi}{2} + 0.01\left(\frac{\pi}{2}\right)$.

12. Let $f(x) = (x-1)(x-2)(x-8)$. Estimate $f(1 + 10 - 4)$ using mean-value theorem with x_T and $x_A = 1 + 10^{-4}$.

CHAPTER 4

A Survey of Simple Methods and Tools

4.1 Introduction

This chapter aims to provide a some very basic examples of approximations, algorithms, and error estimates. These examples serve two purposes: firstly, they will help us build or use some of the methods discussed in subsequent chapters, and secondly, they will facilitate our comprehension of the subject matter. Our objective is to familiarize the student with the wide scope of numerical computations, without confining our focus to a narrow subset of the field. Furthermore, this study demonstrates the utilization of basic approximation techniques such as difference methods for derivative approximation, linear interpolation, and the solution of tridiagonal systems as fundamental components in computational schemes for more complex scenarios. Specifically, these techniques serve as the foundation for Euler's method in solving initial value problems, the trapezoid rule in numerical integration, and the approximate solution of two-point boundary value problems. The objective of this study is to familiarize the reader with the fundamental concepts of numerical methods and analysis through an examination of elementary approaches applied to a wide range of problem domains.

4.2 Horner's Rule and Nested Multiplication

Horner's method is a widely used computational technique for the efficient evaluation and deflation of polynomials. Moreover, it possesses the capability to effectively compute multiple order derivatives of a polynomial, thus making it a frequently employed component in the implementation of Newton's technique. Horner's Rule is based on the concept of rewriting a polynomial in a way that minimizes the number of multiplications and additions needed for evaluation. Horner's Rule offers a more efficient approach compared to the traditional method of individually multiplying each term by the corresponding power of the variable and then adding them together.

Horner's Rule has a significant application in numerical methods for polynomial interpolation and approximation. It is a valuable tool in curve fitting, numerical integration, and solving differential equations. It allows for efficient evaluation of a polynomial at different values, making it particularly useful when fitting a polynomial to a set of data points. Furthermore, it is commonly employed in computer algorithms that deal with polynomial manipulation, such as computer algebra systems and numerical software libraries.

Steps for Using Horner's Rule

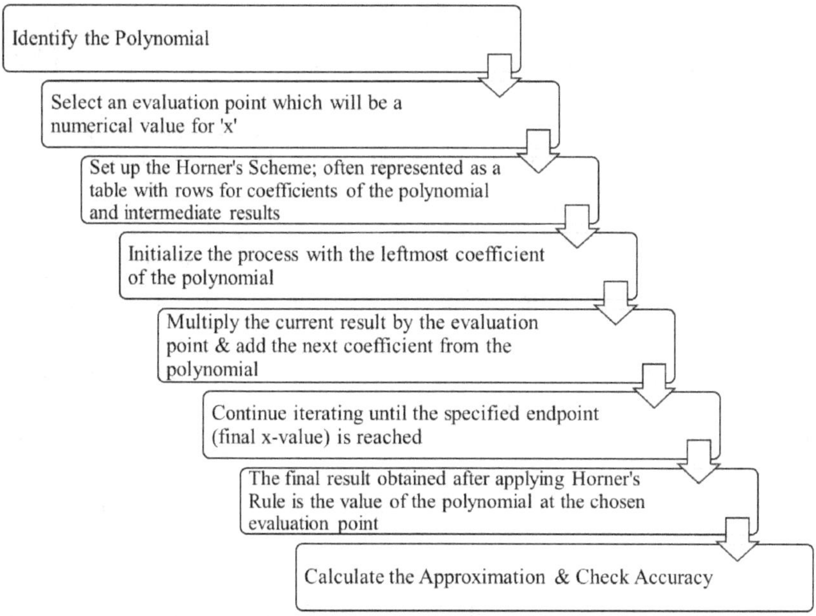

Figure 4.1 Steps for using Hornor's rule

To effectively apply Horner's Rule, it is important to first comprehend the problem at hand and precisely identify the polynomial expression that needs to be evaluated. To construct Horner's scheme, which is typically represented as a table, we begin by considering the coefficient on the far left of the polynomial. In order to perform a series of multiplications and additions, it is important to go through the coefficients from left to right. The determination of the value of a polynomial at a certain evaluation point can be achieved by employing Horner's Rule, resulting in the computation of the ultimate outcome. This method holds significant value in the field of numerical analysis, particularly in tasks pertaining to polynomial interpolation and curve fitting. Efficiently calculating polynomial values is of utmost importance in these tasks, and it plays a crucial role in achieving this.

Consider a real polynomial, denoted as p, which is expressed in standard form. In other words, the coefficients of p are known and are arranged in the standard ordered basis. The sequence consists of the terms $1, x, x^2, \ldots, x^n$, that is, we know the a_k in,

$$p(x) = \sum_{k=0}^{n} a_k x^k \tag{4.1}$$

Or

$$p(x) = a_0 + a_1 x + a_2 x^2 + \ldots + a_n x^n \tag{4.2}$$

Where the variables x^k represent the polynomial variables, while a_k are real values representing the coefficients of the polynomial.

The utilization of Horner's rule in polynomial division is an algorithmic approach aimed at streamlining the evaluation process of a polynomial $p(x)$ at a specific value $x = x_0$. This is achieved by splitting the polynomial into monomials, which are polynomials of the first degree. Each monomial consists of at most one multiplication and one addition operation. The outcome derived from each individual monomial is combined with the outcome derived from the subsequent monomial in a cumulative manner through addition. The process of division referred to in this context is commonly recognized as synthetic division.

In order to provide a comprehensive explanation, let us proceed with rewriting the given polynomial expression in its extended form,

$$p(x_0) = a_0 + a_1 x_0 + a_2 {x_0}^2 + \ldots + a_n {x_0}^n \tag{4.3}$$

Additionally, this can be written as:

$$p(x_0) = a_0 + x_0(a_1 + x_0(a_2 + (x_0(a_3 + \ldots + (a_{n-1} + a_n x_0) \ldots) \tag{4.4}$$

The algorithm suggested by this rule relies on assessing the monomials created above, starting from the one within the innermost parenthesis and progressing outward to evaluate the monomials within the outer parenthesis.

The algorithm is executed by following the steps outlined below:

1. Set $k = n$
2. Let $b_k = a_k$
3. Let $b_{k-1} = a_{k-1} + b_k x_0$
4. Let $k = k - 1$
5. If $k \geq 0$ then go yo step 3
   Else End

The algorithm can also be represented visually by considering the polynomial of degree 5:

$$p(x) = a_0 + a_1 x + a_2 x^2 + a_3 x^3 + a_4 x^4 + a_5 x^5 \tag{4.5}$$

which can be evaluated at $x = x_0$ by re-arranging it as:

$$p(x_0) = a_0 + x_0(a_1 + x_0\left(a_2 + \left(x_0 a_3 + x_0(a_4 + a_5 x_0)\right)\right)) \tag{4.6}$$

Another way of presenting the outcomes using this algorithm is by utilizing the tableau provided in the following way:

Table 4.1 Outcomes Using Horner's Rule Algorithm

k	5	4	3	2	1	0
	$b_5 = a_5$	$b_4 = a_4 + x_0 b_5$	$b_3 = a_3 + x_0 b_4$	$b_2 = a_2 + x_0 b_3$	$b_1 = a_1 + x_0 b_2$	$b_0 = a_0 + x_0 b_1$

Example 4.1 Evaluate the polynomial $f(x) = x^4 + 3x^3 + 5x^2 + 7x + 9$ at $x = 2$

Solution. Given that the polynomial is a quartic, it follows that the value of n is equal to 4.

Chart: 4a

K	4	3	2	1	0
Step	$b_4 = 1$	$b_3 = 3 + 2 \times 1$	$b_2 = 5 + 2 \times 5$	$b_1 = 7 + 2 \times 15$	$b_0 = 9 + 2 \times 37$
Re-sult	1	5	15	37	83

Typically, when assessing a polynomial at a specific value, it is common to substitute that value into the polynomial and perform the necessary calculations. Additionally, it is possible to utilize or create a mathematical computational application for this purpose, as it becomes essential when working with intricate polynomials of significant degrees. By the polynomial remainder theorem, we know that the remainder is $f(2)$. Thus $f(2) = 83$.

Example 4.2 Evaluate $f(x) = 2x^4 - 3x^3 + x^2 - 2x + 3$ when $x = 2$

Solution. Given that the polynomial is a quartic, it follows that the value of n is equal to 4.

Chart: 4b

K	4	3	2	1	0
Ste p	$b_4 = 2$	$b_3 = (-3) + 2 \times 2$	$b_2 = 1 + 2 \times 1$	$b_1 = (-2) + 2 \times 3$	$b_0 = 3 + 2 \times 4$
Re-sult	2	1	3	4	1 1

Be careful when writing Homer's rule for a polynomial which has some coefficients that are negative. It is best to simply store the coefficients as negative numbers rather than try to (correctly) distribute the negative signs throughout the nested multiplication. Thus $f(2) = 11$.

4.3 Difference Approximations to the Derivative

A finite difference approach is a method that uses the function at different points to obtain equivalent to an ordinary or partial derivative.

You can get a good idea of the answer to an ordinary or partial differential equation by getting a good idea of the derivatives at a set of grid points that are usually close to each other.

One of the fundamental applications of Taylor's Theorem in the field of creating approximations is the utilization of difference quotients to estimate the derivative of a given function. This can be simply understood from the formal definition of the derivative:

$$f'(x) = \lim_{h\to 0}\frac{f(x+h)-f(x)}{h} \Rightarrow f'(x) \approx \frac{f(x+h)-f(x)}{h}$$

The accuracy of the approximation can be evaluated using a simple computation applying Taylor's Theorem:

$$f'(x) - \frac{f(x+h)-f(x)}{h} = f'(x) - \frac{hf'(x)+\frac{1}{2}h^2f''(\xi_{x,h})}{h}$$
$$= -\frac{1}{2}hf''(\xi_{x,h})$$

So that we have,

$$f'(x) - \frac{f(x+h)-f(x)}{h} = --\frac{1}{2}hf''(\xi_{x,h}) = O(h) \qquad (4.7)$$

In this case, the error is proportional to the parameter h.

It is possible to improve upon our current situation. The enhancement does not require a significant amount of effort. Let us consider the two Taylor expansions:

$$f(x+h) = f(x) + hf'(x) + \frac{1}{2}h^2f''(x) + \frac{1}{6}h^3f'''(\xi_1) \qquad (4.8)$$

And

$$f(x-h) = f(x) - hf'(x) + \frac{1}{2}h^2f''(x) + \frac{1}{6}h^3f'''(\xi_2) \qquad (4.9)$$

Now subtracting these to get

$$f(x+h) - f(x-h) = 2hf'(x) + \frac{1}{6}h^3f'''(\xi_1) + +\frac{1}{6}h^3f'''(\xi_2)$$

solving this for $f'(x)$ to get

$$f'(x) = \frac{f(x+h)-f(x-h)}{2h} - \frac{1}{6}h^2\frac{f'''(\xi_1)-f'''(\xi_2)}{2} \tag{4.10}$$

or,
we can use the Discrete Average Value Theorem,

$$f'(x) = \frac{f(x+h)-f(x-h)}{2h} - \frac{1}{6}h^2 f''(\xi_{x,h}) \tag{4.11}$$

The variable $\xi_{x,h}$ is dependent on both x and h. It is valuable to undertake a comparison between the two approaches and their respective error estimates. It is important to highlight that the computational costs of the two approximations are almost equal. Both of these methods necessitate the evaluation of two functions and only a small number of additional arithmetic operations. Given the arbitrary nature of the function, it is plausible that its computational cost may be substantial. Consequently, we consider this aspect to be the most prominent component of the overall cost. The initial approximation has an error that is depending on f'' and indicates a direct proportionality to h. Conversely, the subsequent approximation is contingent upon f''' and exhibits a direct proportionality to h^2. It is evident that in order to achieve optimal accuracy, it is advisable to utilize a smaller value for h. Consequently, the second estimate is expected to be superior, as the significance of $h^2 < h$ outweighs the impact of the varying values of f'' and f'''.

Table 4.2 Forward Finite-Difference Formulas

First Derivative:	Error
$f'(x_i) = \frac{f(x_{i+1}) - f(x_i)}{h}$	$O(h)$
$f'(x_i) = \frac{-f(x_{i+2}) + 4f(x_{i+1}) - 3f(x_i)}{2h}$	$O(h^2)$
Second Derivative:	
$f''(x_i) = \frac{f(x_{i+2}) - 2f(x_{i+1}) + f(x_i)}{h^2}$	$O(h)$
$f''(x_i) = \frac{-f(x_{i+3}) + 4f(x_{i+2}) - 5f(x_{i+1}) + 2f(x_i)}{h^2}$	$O(h^2)$
Third Derivative:	
$f'''(x_i) = \frac{-f(x_{i+3}) - 3f(x_{i+2}) + 3f(x_{i+1}) - f(x_i)}{h^3}$	$O(h)$
$f'''(x_i) = \frac{-3f(x_{i+4}) - 14f(x_{i+3}) - 24f(x_{i+2}) + 18f(x_{i+1}) - 5f(x_i)}{2h^3}$	$O(h^2)$

Fourth Derivative:	
$f''''(x_i) = \frac{f(x_{i+4}) - 4f(x_{i+3}) + 6f(x_{i+2}) - 4f(x_{i+1}) + f(x_i)}{h^4}$	$O(h)$
$f''''(x_i) = \frac{-2f(x_{i+5}) + 11f(x_{i+4}) - 24f(x_{i+3}) + 26f(x_{i+2}) - 14f(x_{i+1}) + 3f(x_i)}{h^4}$	$O(h^2)$

The backward finite-difference formulas can be used to get close to derivatives in repeated numerical formulas.

Table 4.3 Backward Finite-Difference Formulas

First Derivative:	Error
$f'(x_i) = \frac{f(x_i) - f(x_{i-1})}{h}$	$O(h)$
$f'(x_i) = \frac{-3f(x_i) - 4f(x_{i-1}) + f(x_{i-2})}{2h}$	$O(h^2)$
Second Derivative:	
$f''(x_i) = \frac{f(x_i) - 2f(x_{i-1}) + f(x_{i-2})}{h^2}$	$O(h)$
$f''(x_i) = \frac{2f(x_i) - 5f(x_{i-1}) + 4f(x_{i-2}) - f(x_{i-3})}{h^2}$	$O(h^2)$
Third Derivative:	
$f'''(x_i) = \frac{f(x_i) - 3f(x_{i-1}) + 3f(x_{i-2}) - f(x_{i-3})}{h^3}$	$O(h)$
$f'''(x_i) = \frac{5f(x_i) - 18f(x_{i-1}) + 24f(x_{i-2}) - 14f(x_{i-3}) + -3f(x_{i-4})}{2h^3}$	$O(h^2)$
Fourth Derivative:	
$f''''(x_i) = \frac{f(x_i) - 4f(x_{i-1}) + 6f(x_{i-2}) - 4f(x_{i-3}) + f(x_{i-4})}{h^4}$	$O(h)$
$f''''(x_i) = \frac{3f(x_i) - 14f(x_{i-1}) + 26f(x_{i-2}) - 24f(x_{i-3}) + 11f(x_{i-4}) - 2f(x_{i-5})}{h^4}$	$O(h^2)$

Table 4.4 Centered Finite-Difference Formulas

First Derivative:	Error
$f'(x_i) = \dfrac{f(x_{i+1}) - f(x_{i-1})}{2h}$	$O(h^2)$
$f'(x_i) = \dfrac{-f(x_{i+2}) + 8f(x_{i+1}) - 8f(x_{i-1}) + f(x_{i-2})}{12h}$	$O(h^4)$
Second Derivative:	
$f''(x_i) = \dfrac{f(x_{i+1}) - 2f(x_i) + f(x_{i-1})}{h^2}$	$O(h^2)$
$f''(x_i) = \dfrac{-f(x_{i+2}) + 16f(x_{i+1}) - 30f(x_i) + 16f(x_{i-1}) - f(x_{i-2})}{12h^2}$	$O(h^4)$
Third Derivative:	
$f'''(x_i) = \dfrac{f(x_{i+2}) - 2f(x_{i+1}) + 2f(x_{i-1}) - f(x_{i-2})}{2h^3}$	$O(h^2)$
$f'''(x_i) = \dfrac{-f(x_{i+3}) + 8f(x_{i+2}) - 13f(x_{i+1}) + 13f(x_{i-1}) - 8f(x_{i-2}) + f(x_{i-3})}{8h^3}$	$O(h^4)$
Fourth Derivative:	
$f''''(x_i) = \dfrac{f(x_{i+2}) - 4f(x_{i+1}) + 6f(x_i) - 4f(x_{i-1}) + f(x_{i-2})}{h^4}$	$O(h^2)$
$f''''(x_i) = \dfrac{-f(x_{i+3}) + 12f(x_{i+2}) + 39f(x_{i+1}) + 56f(x_i) - 39f(x_{i-1}) + 12f(x_{i-2}) + f(x_{i+3})}{6h^4}$	$O(h^4)$

Example 4.3 Estimate the first derivative of

$$f(x) = -0.1x^4 - 0.15x^3 - 0.5x^2 - 0.25x + 1.2$$

At $x = 0.5$ using finite differences and a step size $h = 0.25$. Employ the $O(h)$ and $O(h^2)$ forward and backward difference formulas, and the $O(h^2)$ and $O(h^4)$. centered difference formulas. Compute the absolute relative true error (in percent).

Solution. First, let us compute the exact derivative by direct differentiation.

$$f'(x) = -0.4x^3 - 0.45x^2 - x - 0.25$$

Then, by substituting $x = 0.5$ we obtain the true value $f'(0.5) = -0.9125$.

Now, for $h = 0.25$ and $x_i = 0.5$, we obtain the necessary data:

$$x_{i-2} = x_i - 2h = 0.5 - 2(0.25) = 0; \qquad f(x_{i-2}) = f(0) = 1.2$$
$$x_{i-1} = x_i - h = 0.5 - 0.25 = 0.25; \qquad f(x_{i-1}) = f(0.25)$$
$$= 1.1035156$$
$$x_i = 0.5; \qquad f(x_i) = 0.925$$
$$x_{i+1} = x_i + h = 0.5 + 0.25 = 0.75; \qquad f(x_{i+1}) = f(0.75)$$
$$= 0.6363281$$
$$x_{i+2} = x_i + 2h = 0.5 + 0.5 = 1; \qquad f(x_{i+2}) = f(1) = 0.2$$

which we may use to compute the $\mathcal{O}(h)$ forward difference (and relative error)

$$f'(0.5) \cong \frac{0.6363281 - 0.925}{0.25} = -1.155; \qquad (\varepsilon = 26.6\%)$$

The $\mathcal{O}(h^2)$ forward difference,

$$f'(0.5) \cong \frac{-0.2 + 4(0.6363281) - 3(0.925)}{2(0.25)}$$
$$= -0.859375; \qquad (\varepsilon = 5.82\%)$$

The $\mathcal{O}(h)$ backward difference,

$$f'(0.5) \cong \frac{0.935 - 1.1035156}{0.25} = -0.714; \qquad (\varepsilon = 21.7\%)$$

The $\mathcal{O}(h^2)$ backward difference,

$$f'(0.5) \cong \frac{3(0.925) - 4(1.1035156) + 1.2}{2(0.25)}$$
$$= -0..878125; \qquad (\varepsilon = 3.77\%)$$

The $\mathcal{O}(h^2)$ centered difference,

$$f'(0.5) \cong \frac{0.6363281 - 1.1035156}{2(0.25)} = -0.934; \qquad (\varepsilon = 2.4\%)$$

The $\mathcal{O}(h^4)$ centered difference,

$$f'(0.5) \cong \frac{-0.2 + 8(0.6363281) - 8(1.1035156) + 1.2}{12(0.25)}$$
$$= -0..878125; \qquad (\varepsilon = 0\%)$$

As expected, the estimates derived from the $\mathcal{O}(h^2)$ formulas indicate significantly higher accuracy compared to their $\mathcal{O}(h)$ counterparts. The formula for center difference $\mathcal{O}(h^4)$ is accurate for polynomials of fourth-order or lower. The reason for this is that the formula, which is based on the Taylor series, utilizes four points to approximate the derivative of $f(x)$, denoted as $f'(x)$. By employing four points, we are able to precisely interpolate a cubic polynomial that represents the derivative $f'(x)$ of a fourth-degree function $f(x)$.

Another method to provide justification for the precise nature of the $\mathcal{O}(h^4)$ central difference formula for certain functions is by acknowledging that the truncation error can be expressed as

$$E_t = \frac{h^4 f^5(\gamma)}{5!} \tag{4.12}$$

If $f(x)$ represents a polynomial of degree 4, then $f^5(x) = 0$, for all x, resulting in $E_t = 0$.

Therefore, the appearance of the convergence rate column in the previous tables enables us to determine the category of functions for which the corresponding derivative formula is precise. If the error is of order $\mathcal{O}(h^n)$ then, the derivative formula is exact for all cases. The polynomial function $f(x)$ is defined up to and including degree n. Chapter 8 contains more material on the approximation of definite integrals.

4.4 The Trapezoid Rule: Application

The trapezoid rule for estimating definite integrals is one of the most important ways that linear interpolation is used. The trapezoidal rule utilizes a first-degree Lagrange approximating polynomial with nodes and linear interpolation. It is used to approximate the definite integral of a function over a given interval. On the other hand, polynomial interpolation involves approximating functions by fitting polynomials to data points. Combining these methods can be helpful when dealing with functions that are difficult to integrate analytically. The Trapezoidal Rule can be used to estimate the integral of a polynomial representation of a function, which is created by applying polynomial interpolation to discrete data points. This numerical solution for integration is obtained by using the Trapezoidal Rule on the polynomial. The integration of interest can be defined as the integral of a function, denoted as $I(f)$, thus:

$$I(f) = \int_a^b f(x)dx$$

Let $p_1(x)$ be the linear polynomial that interpolates f at $x = a$ and $x = b$:

$$p_1(x) = \frac{x-a}{b-a} f(b) + \frac{b-x}{b-a} f(a)$$

Then the basic trapezoid rule is defined by exactly integrating $p_1(x)$:

$$T_1(f) = I(p_1) = \int_a^b p_1(x)dx = \frac{1}{2}(b-a)\big(f(b) + f(a)\big) \qquad (4.13)$$

In this context, we are utilizing an approximation method by substituting the "exact function" (f) with a simpler function (p_1) that provides a close approximation. We then perform the desired calculation (integration) precisely on the simpler function.

Figure 4.2 depicts the aforementioned approximation. The expression to be integrated is the absolute value of the sum of sine of $\frac{1}{2} + sin\, \pi x$ times the variable χ, over the given interval $[\frac{1}{2}, 1]$.

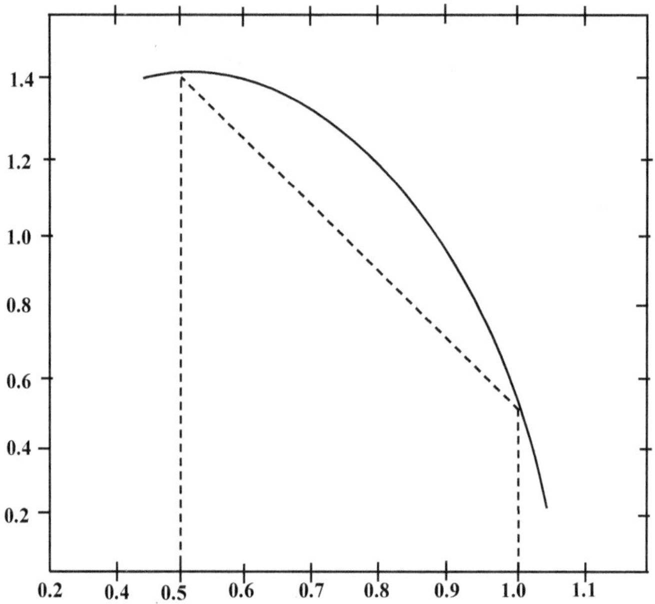

Figure 0.2 Illustration of the basic trapezoid rule

The illustration indicates that the error, which is $I(f) - T_1(f)$, may be substantial. We are able to analyze the error due to the construction of T_1 as the exact integral of p_1. This allows us to apply the interpolation error theory discussed in Eq. 4.13.

$$\begin{aligned} I(f) - T_1(f) &= I(f) - I(p_1) \\ &= \int_a^b f(x)dx - \int_a^b f(p_1)dx \\ &= \int_a^b (f(x)dx - \int_a^b p_1(x))dx \\ &= \frac{1}{2}\int_a^b (x-a)(x-b)f''(\xi_x)dx \end{aligned}$$

Where, $\xi_x \in [a,b]$ depends on x. Since the expression $(x — a)(x — b)$ is unchanged in sign over the interval [a, 6], we can proceed to employ the Integral Mean Value Theorem in order to obtain an estimation of the inaccuracy.

$$\frac{1}{2}\int_a^b (x-a)(x-b)f''(\xi_x)dx = f''(\eta)\int_a^b (x-a)(x-b)dx$$
$$= -\frac{1}{6}(b-a)^3 f''(\eta)$$

So that we have,

$$I(f) - T_1(f) = -\tfrac{1}{12}(b-a)^3 f''(\eta) \tag{4.14}$$

Where, $\eta \in [a, b]$.

Accordingly, we have established the following theorem:

Theorem 4.1 (Trapezoid Rule Error Estimate, Single Subinterval)

Let f
$\in C^2([a, b])$ *and let* p_1 *interpolate* f *at* a *and* b. *Define* $T_1(f)$
$= I(p_1)$.
Then there exists $\eta \in [a, b]$ *such that*

$$I(f) - T_1(f) = -\tfrac{1}{12}(b-a)^3 f''(\eta) \tag{4.15}$$

While this outcome is correct, its practical utility seems limited as the error will only be negligible when the integration interval, denoted as $b - a$, is tiny. Additionally, it should be noted that this outcome represents an initial stage in the process. It is important to employ a greater number of data points in the process of approximation at now. In order to achieve this objective, the interval $[a, b]$ is partitioned into n subintervals by means of meshpoints x_i, where $0 < i < n$:

$$a = x_0 < x_1 < x_2 \ . \ . \ . < x_{n-1} < x_n = b \tag{4.16}$$

The basic trapezoid rule (4.13) is applied over each subinterval $[x_{i-1}, x_i]$ to obtain the composite trapezoid rule, which is commonly known as "the n-subinterval trapezoid rule" or simply the trapezoid rule when there is no chance of confusion.

$$I(f) = \sum_{i=1}^{n} \int_{x_{i-1}}^{x_i} f(x)dx$$

$$\approx \sum_{i=1}^{n} \frac{1}{2}(x_i - x_{i-1})(f(x_{i-1}) + f(x_i))$$
$$= T_n(f)$$

If we utilize a uniform grid, where the mesh points are evenly spaced with a distance of h between x_i and x_{i-1} , so that, $x_i - x_{i-1} = h$ for all i, this greatly simplifies the situation:

$$T_n(f) = \frac{h}{2}(f(x_0) + 2f(x_1) + 2f(x_2) + \ldots + 2f(x_{n-1}) + f(x_n)) \tag{4.17}$$

Figure 4.3 illustrates the application of the same integral of $f(x) = \frac{1}{2} + \sin \pi x$; as shown in Figure 4.2. It is important to note that even with a minimal number of additional points in the grid, the apparent error is significantly reduced compared to the previous case. What we have essentially done here is perform an exact integration of a piecewise linear interpolation of the function f.

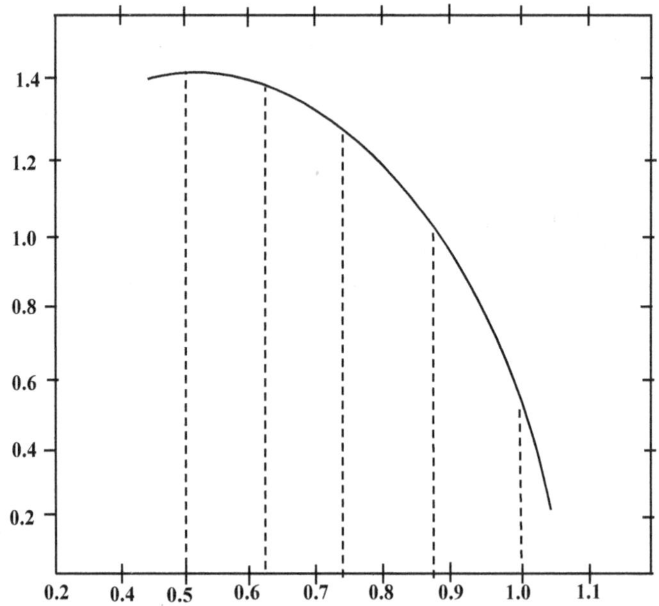

Figure 0.3 Illustration of the trapezoid rule

The accuracy of the n-subinterval rule can be easily determined based on the previous result.

Theorem 4.2 (Trapezoid Rule Error Estimate, Uniform Grid)

Let $f \in C^2([a,b])$ and let $T_n(f)$ be the n −subinterval trapezoid rule approximation to $I(f)$, using a uniform grid. Then there exists $\xi_h \in [a,b]$, depending on h, such that

$$I(f) - T_n(f) = -\frac{b-a}{12} h^2 f''(\xi_h) \tag{4.18}$$

This theorem provides valuable insights into the n-subinterval trapezoid rule. First, it is stated that the numerical approximation will converge to the exact value as long as the function is continuous and differentiable in the interval $f \in C^2([a,b])$:

$$\begin{aligned}\lim_{n\to\infty} |I(f) - T_n(f)| &= \lim_{h\to 0} \frac{b-a}{12} h^2 |f''(\xi_h)| \\ &\leq \left(\lim_{h\to 0} \frac{b-a}{12} h^2\right)\left(\max_{x\in[a,b]} |f''|\right) = 0\end{aligned}$$

So that,

$$\lim_{n\to\infty} T_n(f) = I(f)$$

Additionally, it informs us about the rate at which this convergence occurs, as the error is proportional to h^2. The easiest way to understand this is by observing that when the number of subintervals is doubled, the error decreases by approximately a factor of 4. We use the term "roughly" because the value of ξ_h is dependent on h and will slightly vary as we adjust the mesh size.

It is important to highlight that the error estimate's value lies not in our ability to predict the error beforehand, although there are instances where we can use the error theory to estimate it. Rather, the error estimate informs us about the rate at which the error will decrease as the mesh size, h, changes. This information will be beneficial in Chapter 5 as we aim to enhance the trapezoid rule.

An example will help to illustrate the point.

Example 4.4 Let $f(x) = e^x$ and $[a,b] = [0,1]$ so that $I(f) = e - 1 = 1.71828 \ldots$ Then the trapezoid rule using a single subinterval is given by

$$T_1(f) = \frac{1}{2} \times 1 \times (e^0 + e^1) = \frac{(1+e)}{2} = 1.859140914$$

Whereas the trapezoid rule using two subintervals is given by

$$T_2(f) = \frac{1}{2} \times \frac{1}{2} \times \left(e^0 + 2e^{\frac{1}{2}} + e^1\right) = \frac{1}{4}(1 + 2\sqrt{2} + e)$$
$$= 1.753931093$$

It is observed that the second value exhibits a significantly higher level of accuracy, as anticipated.

If we proceed with the calculation by employing a series of uniformly spaced grids with $h^{-1} = 2,4,8,\ldots,2048,$ the results presented in Table 4-5 are obtained. It is important to observe that the final column, which presents the ratio between the old error and the current error, clearly demonstrates a reduction in error by about a factor of 4 when the number of points is doubled.

Table 4.5 Trapezoid example, $f(x) = e, [a,b] = [0]$

n	$T_n(f)$	$I(f) - T_n(f)$	Error Ratio
2	1.753931092	$-0.356493\,E-01$	N/A
4	1.727221905	$-0.894008\,E-02$	3.9876
8	1.720518592	$-0.223676\,E-02$	3.9969
16	1.718841129	$-0.559300\,E-03$	3.9992
32	1.718421660	$-0.139832\,E-03$	3.9998
64	1.718316787	$-0.349584\,E-04$	4.0000
128	1.718290568	$-0.873962E-05$	4.0000
256	1.718284013	$-0.218491\,E-05$	4.0000
512	1.718282375	$-0.546227E-06$	4.0000
1024	1.718281965	$-0.136557\,E-06$	4.0000
2048	1.718281965	$-0.341392\,E-07$	4.0000

Programming Hint: It is important to understand that using the method of increasing the number of subintervals and monitoring the subsequent decrease in error by around a factor of 4 can serve as a valuable tool for troubleshooting programs utilizing the trapezoid rule. If the software does not exhibit mistakes with such behavior, then there are only three potential inferences to be inferred. There are three possible explanations for the observed behavior:

1. The test function does not meet the condition of being continuously differentiable twice C^2 on the entire interval of integration;
2. There is an error in the code implementation;
3. The example is related to a specific category of functions, such as periodic functions integrated over an integer multiple of the period, for which the trapezoid rule exhibits exceptional accuracy. In this case, the error initially starts at a very small value and may not consistently decrease in a predictable manner. One should refrain from accepting results that don't show a proper decrease in the rate of error, unless one can provide evidence that either the initial or third possibility is applicable to the given example:

Example 4.5 The error theory can sometimes be used to predict how small the mesh must be to achieve a specified desired error. For example, consider the integral

$$I(f) = \int_0^1 e^{-x^2} dx$$

How small does h have to be to guarantee that $|I(f) - T_n(f)| \leq 10^{-3}$?

Solution: We know that

$$|I(f) - T_n(f)| = \frac{1}{12} h^2 |f''(\xi_x)| \leq \frac{1}{12} h^2 \max_{x \in [0,1]} |f''(x)|$$

Ordinary calculus shows us that

$$f''(x) = e^{-x^2}(4x^2 - 2)$$

so that, for all $x \in [0,1]$,

$$|f''(x)| = \left|e^{-x^2}(4x^2 - 2)\right| \leq e^0(4 + 2) = 6$$

Therefore,

$$|I(f) - T_n(f)| \leq 10^{-3} \Leftarrow \frac{1}{12} h^2(6) \leq 10^{-3}$$

which implies that we need $h \leq \sqrt{0.002} = 0.044721 \ldots$ The difficult part of this computation lies in obtaining an upper bound for the second derivative factor in the error estimate. The assumption of a uniform grid can be helpful, but it is not necessary.

4.5 Solution of Tridiagonal Linear Systems

In computer mathematics, one of the most important things to learn is how to solve linear systems of equations. We will not be providing an in-depth presentation on this topic as it requests an entirely separate course. This section focuses on a specific subject of interest: the solution of tridiagonal systems of linear equations.

$$Ax = b \tag{4.19}$$

where x is an unknown n-vector, b is the given n-vector, and A is a tridiagonal $n \times n$ matrix.

$$A = \begin{bmatrix} a_{11} & a_{12} & 0 & \cdots & 0 \\ a_{21} & a_{22} & a_{23} & \cdots & 0 \\ 0 & a_{32} & a_{33} & \cdots & 0 \\ \vdots & \vdots & \vdots & \ddots & a_{n-1,n} \\ 0 & 0 & 0 & a_{n,n-1} & a_{nn} \end{bmatrix}$$

Before starting the method, it is necessary that we undertake a notational simplification. Instead of employing the conventional matrix notation and indexing style, we shall utilize the symbols l_i, d_i *and* u_i to represent the lower-diagonal, diagonal, and upper-diagonal components, respectively.

$$\begin{aligned} l_i &= a_{i,i-1}, & 2 \leq i \leq n, \\ d_i &= a_{ii}, & 1 \leq i \leq n, \\ u_i &= a_{i,i+1}, & 1 \leq i \leq n-1, \end{aligned}$$

where it is assumed that the convention that $l_i = 0$ and $u_n = 0$. Under this notation, the augmented matrix that corresponds to the system is

$$[A|f] = \left[\begin{array}{ccccc|c} d_1 & u_1 & 0 & \cdots & 0 & f_1 \\ l_2 & d_2 & u_2 & \cdots & 0 & f_2 \\ 0 & l_3 & d_3 & \cdots & 0 & \vdots \\ \vdots & \vdots & \vdots & \ddots & u_{n-1} & f_{n-1} \\ 0 & 0 & 0 & l_n & d_n & f_n \end{array}\right] \tag{4.20}$$

When saving the problem, we can use only the four vectors for l, d, u, and f instead of the entire n-by-n matrix, as the matrix mostly consists of zeroes.

By using the elimination process, the augmented matrix is reduced to an equivalent upper triangular form:

$$[T|g] = \left[\begin{matrix} \delta_1 & u_1 & 0 & \cdots & 0 \\ l_2 & \delta_2 & u_2 & \cdots & 0 \\ 0 & l_3 & \delta_3 & \cdots & 0 \\ \vdots & \vdots & \vdots & \ddots & u_{n-1} \\ 0 & 0 & 0 & l_n & d_n \end{matrix}\right| \left.\begin{matrix} g_1 \\ g_2 \\ \vdots \\ g_{n-1} \\ g_n \end{matrix}\right]$$

Where, $\delta_1 = d_1, \ \delta_2 = d_2 - u_1\left({}^{l_2}/_{\delta_1}\right), \ \delta_3 = d_3 - u_2\left({}^{l_3}/_{\delta_2}\right)$

And generally, $\delta_k = d_k - u_{k-1}\left({}^{l_k}/_{\delta_{k-1}}\right), \ 2 \le k \le n$

Similarly,

$$g_1 = f_1, g_2 = f_2 - g_1\left({}^{l_2}/_{\delta_1}\right), \ g_3 = f_3 - g_2\left({}^{l_3}/_{\delta_2}\right),$$

so that the general form is

$$g_k = f_k - g_{k-1}\left({}^{l_k}/_{\delta_{k-1}}\right), \qquad 2 \le k \le n$$

The matrix $[T|g]$ shows row equivalence to the initial augmented matrix $[A|f]$, indicating that the two matrices may be transformed into one another using elementary row operations. Consequently, both augmented matrices correspond to systems that possess identical solution sets. Furthermore, the answer can be readily obtained by solving the final equation, $\delta_n x_n = g_n$ to get $x_n = {}^{g_n}/_{\delta_n}$. Subsequently, this value can be utilized in the preceding equation to determine x_{n-1} and so forth, enabling the determination of each solution component. Once again, this part of the calculation needs to be based on the concept that each $\delta_k \neq 0, 1 \le k \le n$.

Example 4.6 Consider the system

$$\begin{bmatrix} 4 & 1 & 0 & 0 \\ 1 & 4 & 1 & 0 \\ 0 & 1 & 4 & 1 \\ 0 & 0 & 1 & 4 \end{bmatrix} \begin{bmatrix} x_1 \\ x_2 \\ x_3 \\ x_4 \end{bmatrix} = \begin{bmatrix} 9 \\ 12 \\ 18 \\ 19 \end{bmatrix}$$

If we do the elimination stage of the algorithm, we get the new system (written now as a single augmented matrix):

$$[T|g] = \left[\begin{array}{cccc|c} 4 & 1 & 0 & 0 & 6 \\ 0 & \frac{15}{4} & 1 & 0 & \frac{21}{2} \\ 0 & 0 & \frac{56}{15} & 1 & \frac{76}{5} \\ 0 & 0 & 0 & \frac{209}{56} & \frac{209}{14} \end{array}\right]$$

from which we get the solution

$$x_1 = 1, x_2 = 2, x_3 = 3, x_4 = 4$$

On the other hand, if we look at the system

$$\begin{bmatrix} 4 & 1 & 0 & 0 \\ 1 & 4 & 1 & 0 \\ 0 & 1 & 4 & 1 \\ 0 & 0 & 1 & 4 \end{bmatrix} \begin{bmatrix} x_1 \\ x_2 \\ x_3 \\ x_4 \end{bmatrix} = \begin{bmatrix} 9 \\ 12 \\ 18 \\ 19 \end{bmatrix}$$

then the elimination stage produces the following augmented matrix after a single pass through the first loop:

$$[T|g] = \left[\begin{array}{cccc|c} 4 & 1 & 0 & 0 & 9 \\ 1 & 4 & 1 & 0 & 3 \\ 0 & 1 & 4 & 1 & 18 \\ 0 & 0 & 1 & 4 & 19 \end{array}\right]$$

It is evident that we cannot proceed with the process because the next step would involve dividing by zero. However, it is important to note that the fact that the algorithm fails to work on this example does not necessarily imply that the system in question does not have a solution. The reader should ensure that

$$x_1 = 1, x_2 = 2, x_3 = 3, x_4 = 4$$

is indeed the solution to the above system.

Definition 4.1 (Diagonal Dominance for Tridiagonal Matrices)

A tridiagonal matrix is called diagonally dominant if

$$d_i > |l_i| + |u_i| > 0, \ 1 \le i \le n \tag{4.21}$$

For example, the matrix

$$A_1 = \begin{bmatrix} 6 & 1 & 0 & 0 \\ 2 & 6 & 3 & 0 \\ 0 & 6 & 9 & 0 \\ 0 & 0 & 3 & 4 \end{bmatrix}$$

is diagonally dominant, but

$$A_2 = \begin{bmatrix} 6 & 1 & 0 & 0 \\ 2 & 6 & 3 & 0 \\ 0 & 6 & 9 & 0 \\ 0 & 0 & 3 & 4 \end{bmatrix}$$

is not. (The inequality does not hold in the second row.)

If the matrix is diagonally dominant, we can demonstrate that there is no need to be concerned about the diagonal elements becoming zero during the elimination process.

4.6 Simple Two-Point Boundary Value Problems: Application

Consider the problem of finding a function $u(x)$ such that

$$-u'' + u = f(x), \quad x \in [0,1] \tag{4.22}$$

$$u(0) = u(1) = 0 \tag{4.23}$$

Here, the function f is already known, and we are looking for a function u that satisfies the given differential equation and the conditions at $x = 0$ and $x = 1$. However, we currently have enough tools to compute solutions, although a complete analysis of the error is not possible. Problems of such type commonly arise in various fields of applied mathematics and engineering, with structural mechanics being a particularly notable example.

If the interval [0,1] is divided into n equal subintervals $[x_{k-i}, x_k]$, according to

$$0 = x_0 < x_1 < x_2 <. . .< x_{n-1} < x_n = 1$$

With $x_k - x_{k-1} = h$ (therefore $x_k = kh$), and let $U_h(x)$ denote the approximation to $u(x)$. We can use Taylor expansions to derive an approximation to the second derivative, by adding them (instead of subtracting them). After a little manipulation we find that

$$u''(x) - \frac{u(x-h) - 2u(x) + u(x+h)}{h^2} = \frac{1}{12}h^2u^{(4)}(\eta_x) \tag{4.24}$$

For some $\eta_x \in [x — h, x + h]$.. Therefore, the differential equation $-u'' + u = f(x)$ implies that

$$\frac{-u(x-h) - 2u(x) + u(x+h)}{h^2} + u(x) = f(x) + \frac{1}{12}h^2u^{(4)}(\eta_x)$$

To define our approximation, we drop the remainder term and replace u with U_h, which yields

$$\frac{-U_h(x-h) - 2U_h(x) + U_h(x+h)}{h^2} + U_h(x) = f(x) \tag{4.25}$$

This holds for all $x \in [0,1]$. To get a practical computational problem, we impose (4.25) only on our grid points, i.e., we seek the $n — 1$ values $U_k = U_h(x_k)$, $1 \le k \le n-1$, which satisfy (recall that the boundary conditions will force $U_0 = U_n = 0$)

$$-U_{k-1} + (2 + h^2)U_k - U_{k+1} = h^2 f(x_k), \qquad 1 \le k \le n-1$$

The above system of equations can be classified as a tridiagonal system. When expressed in the form of a matrix-vector equation, the equation can be written as follows:

$$\begin{bmatrix} 2+h^2 & -1 & 0 & \cdots & \cdots & 0 \\ -1 & 2+h^2 & -1 & 0 & \cdots & 0 \\ 0 & -1 & 2+h^2 & -1 & 0 & \vdots \\ \vdots & & \ddots & \ddots & \ddots & 0 \\ \vdots & & & \ddots & \ddots & -1 \\ 0 & \cdots & \cdots & 0 & -1 & 2+h^2 \end{bmatrix} \begin{bmatrix} U_1 \\ U_2 \\ \vdots \\ \vdots \\ U_{n-2} \\ U_{n=1} \end{bmatrix} = \begin{bmatrix} h^2 f(x_1) \\ h^2 f(x_2) \\ \vdots \\ \vdots \\ h^2 f(x_{n-2}) \\ h^2 f(x_{n-1}) \end{bmatrix}$$

Example 4.6 If we take the specific case of $h = \frac{1}{4}$ and $f(x) = 1$, then we have a 3×3 system to solve, that is,

$$\begin{bmatrix} 2.0625 & -1 & 0 \\ -1 & 2.0625 & -1 \\ 0 & -1 & 2.0625 \end{bmatrix} \begin{bmatrix} U_1 \\ U_2 \\ U_3 \end{bmatrix} = \begin{bmatrix} 0625 \\ 0625 \\ 0625 \end{bmatrix}$$

Here, by solving matrixes, we get,

$$\begin{aligned} U_1 &= 0.08492201039861 \\ U_2 &= 0.11265164644714 \\ U_3 &= 0.08492201039861 \end{aligned}$$

Thus we have,

$$u\left(\frac{1}{4}\right) \approx 0.08492201039861$$
$$u\left(\frac{1}{2}\right) \approx 0.11265164644714$$
$$u\left(\frac{1}{3}\right) \approx 0.08492201039861$$

Table 4.10 presents the differences seen between the values of $u(x_k)$ and U_k as the step size h decreases from $1/_4$ to $1/_{1024}$, while maintaining the same selection of $f(x)$. The exact solution is given by

$$u(x) = 1 - \frac{e-1}{e^2-1}e^x - \frac{e^2-e}{e^2-1}e^{-x}$$

It should be noted that the magnitude of the mistake exhibits a reduction of approximately fourfold with each doubling of the number of points in the grid. Figure 4-4 provides a graphical representation of the estimated solution and the precise solution for the situation

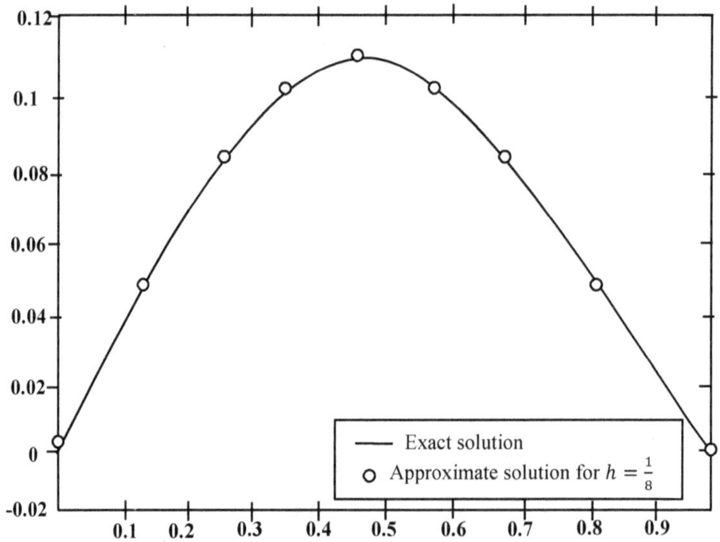

Figure 4.4 Approximate solution to two-point BVP $-u'' + u = 1, for\ h = \frac{1}{8}$

where $h = \frac{1}{8}$. It should be noted that the estimated solution is nearly precise, especially when considering the use of a coarse mesh.

Table 4.6 Two--point boundary value problem results

h^{-1}	$\max \lvert u(x_k) - U_k \rvert$
4	0.529469582786e-03
8	0.133142281084e-03
16	0.333344381367e - 04
32	0.833667068484e - 05
64	0.208435910093e-05
128	0.521101741688e-06
256	0.130276180538e-06
512	0.325690907577e - 07
1024	0.814227994750e-08

4.7 Exercises

1. Write each of the following polynomials in nested form:

a) $x^3 + 3x + 2$;

b) $x^6 + 2x^4 + 4x^2 + 1$;

c) $5x^6 + x^5 + 3x^4 + 3x^3 + x^2 + 1$;

d) $x^2 + 5x + 6$

2. Write each of the following polynomials in nested form, but this time take advantage of the fact that they involve only even powers of x to minimize the computations.

a) $1 + x^2 + \frac{1}{2}x^4 + \frac{1}{6}x^6$

b) $1 - \frac{1}{2}x^2 + \frac{1}{24}x^4$

3. Write each of the following polynomials in nested form:
a) $1 - x + x^2 - x^3$
b) $1 - x^2 + \frac{1}{2}x^4 - \frac{1}{6}x^6$
c) $1 - x + \frac{1}{2}x^4 - \frac{1}{3}x^3 - \frac{1}{4}x^5$

4. Compute, by hand, approximations to $f'(1)$ for each of the following functions, using $h = \frac{1}{16}$ and each of the derivative approximations.
a) $f(x) = \sqrt{x+1}$
b) $f(x) = arc\tan x$
c) $f(x) = \sin \pi x$
d) $f(x) = e^{-x}$
e) $f(x) = lnx$

5. Use the error estimate for the centered difference approximation to the first derivative to prove that this approximation will be exact for any quadratic polynomial.

6. Find coefficients A, B, and C so that
a) $f'(x) = Af(x) + Bf(x+h) + Cf(x+2h) + O(h^2)$
b) $f'(x) = Af(x) + Bf(x-h) + Cf(x-2h) + O(h^2)$

7. Use Taylor's Theorem to show that the approximation

$$f'(x) \approx \frac{8f(x+h) - 8f(x-h) - f(x+2h) + f(x-2h)}{12h}$$

is $O(h^4)$.

8. Use the derivative approximation from Problem 7 to approximate /'(l) for the same functions as in Problem 4. Verify that the expected rate of decrease is observed for the error.

9. Let $f(x) = arc\tan x$. Use the derivative approximation from Problem 7 to approximate $f'(\frac{1}{4}\pi)$ using $h^{-1} = 2,4,8, \ldots$ Try to take h small enough that the rounding error effect begins to dominate the mathematical error. For what value of h does this begin to occur? (You may have to restrict yourself to working in single precision.)

10. Define the following function:

$$f(x) = \ln\left(e^{\sqrt{x^2+1}} \sin(\pi x) + \tan \pi x\right)$$

Compute values of f' over the range $[0, \frac{1}{4}]$ using two methods:

a) Using the centered difference formula from (4.5);
b) Using ordinary calculus to find the formula for f'.

11. Let $f(x) = e^x$, and consider the problem of approximating $f'(1)$, as in the text. Let $D_1(h)$ be the difference approximation in (4.1). Using the appropriate values, compute the new approximations

$$\Delta_1(h) = 2D_1(h) - D_1(2h)$$

and compare these values to the exact derivative value. Are they more or less accurate than the corresponding values of D_1? Try to deduce, from your calculations, how the error depends on h.

12. Let $f(x) = e^x$, and consider the problem of approximating $f'(1)$, as in the text. Let $D_2(h)$ be the difference approximation in (4.5). Using the appropriate values, compute the new approximations

$$\Delta_2(h) = \frac{\left(4D_2(h) - D_2(2h)\right)}{3}$$

and compare these values to the exact derivative value. Are they more or less accurate than the corresponding values of D_2? Try to deduce, from your calculations, how the error depends on h.

13. Use the trapezoid rule with $h = \frac{1}{4}$ to approximate the integral

$$I = \int_0^1 x^3 \, dx = \frac{1}{4}$$

You should find that $T_4 = 17/64$. How small does h have to be for the error to be less than 10^{-3} and 10^{-6}?

14. Use the trapezoid rule and $h = \frac{\pi}{4}$ to approximate the integral

$$I = \int_0^{\pi/2} \sin x \, dx = 1$$

How small does h have to be for the error to be less than 10^{-3} and 10^{-6}?

15. Repeat the above with $h = \frac{\pi}{6}$.
16. Apply the trapezoid rule with $h = \frac{1}{8}$, to approximate the integral

$$I = \int_0^1 \frac{1}{\sqrt{1+x^4}} dx = 0.92703733865069$$

How small does h have to be for the error to be less than 10^{-3} and 10^{-6}?

17. Apply the trapezoid rule with $h = \frac{1}{8}$, to approximate the integral

$$I = \int_0^1 x(1-x^2)dx = \frac{1}{4}$$

How small does h have to be for the error to be less than 10^{-3} and 10^{-6}?

18. Apply the trapezoid rule with $h = \frac{1}{8}$, to approximate the integral

$$I = \int_0^1 \ln(1+x)dx = 2\ln 2 - 1$$

How small does h have to be for the error to be less than 10^{-3} and 10^{-6}?

19. Apply the trapezoid rule with $h = \frac{1}{8}$, to approximate the integral

$$I = \int_0^1 \frac{1}{1+x^3} dx = \frac{1}{3}\ln 2 + \frac{1}{9}\sqrt{3}\pi$$

How small does h have to be for the error to be less than 10^{-3} and 10^{-6}?

20. Express the following systems of equations in the matrix form:

a) $3x + 2y + 4z = 7$

$2x + y + z = 7$
$x + 3y + 5z = 2$

b) $2x - z - 2u = -8$
$y + 2z - u = -1$
$x - y - u = -6$
$-x + 3y - 2u = 7$

21. Compute the solution to the following system of equations:

$$\begin{bmatrix} 4 & 2 & 0 & 0 \\ 1 & 4 & 1 & 0 \\ 0 & 1 & 4 & 1 \\ 0 & 0 & 2 & 4 \end{bmatrix} \begin{bmatrix} x_1 \\ x_2 \\ x_3 \\ x_4 \end{bmatrix} = \begin{bmatrix} \pi/9 \\ \sqrt{3}/2 \\ \sqrt{3}/2 \\ -\pi/9 \end{bmatrix}$$

22. Solve the following system of equations:

$$\begin{bmatrix} 6 & 1 & 0 & 0 \\ 2 & 4 & 1 & 0 \\ 0 & 1 & 4 & 2 \\ 0 & 0 & 1 & 6 \end{bmatrix} \begin{bmatrix} x_1 \\ x_2 \\ x_3 \\ x_4 \end{bmatrix} = \begin{bmatrix} 8 \\ 13 \\ 22 \\ 27 \end{bmatrix}$$

You should get the solution $x = (1,2,3,4)^T$

23. Apply the trapezoid rule to the integral

$$I = \int_0^1 \sqrt{x}\, dx = \frac{2}{3}$$

using a sequence of uniform grids with $h = \frac{1}{2}, \frac{1}{4}, \ldots$ Do we get the expected rate of convergence? Explain.

24. Solve, by hand, the two-point BVP when $f(x) = x$, using $h = \frac{1}{4}$. Write out the linear system explicitly prior to solution. You should get the following 3 x 3 system:

$$\begin{bmatrix} 2.0625 & -1 & 0 \\ -1 & 2.0625 & -1 \\ 0 & -1 & 2.0625 \end{bmatrix} \begin{bmatrix} U_1 \\ U_2 \\ U_3 \end{bmatrix} = \begin{bmatrix} 0.015625 \\ 0.03125 \\ 0.046875 \end{bmatrix}$$

25. Repeat the above, this time using $h = \frac{1}{5}$. What is the system now?

26. Write a program that solves the two-point BVP, where f is as given below.

a) $f(x) = 4e^{-x} - 4xe^{-x}$, $\quad u(x) = x(1-x)e^{-x}$

b) $f(x) = (\pi^2 + 1)\sin \pi x$, $\quad u(x) = \sin \pi x$

27. Solve the two-point BVP problem

a) $-u'' + 64u' + u = 1,\ x \in [0,1]$

b) $u(0) = u(1) = 0$,

using a range of mesh sizes, starting with $h = \frac{1}{4}, \frac{1}{8}$, and going as far as $h = \frac{1}{256}$. Comment on your results.

CHAPTER 5

Numerical Solutions of Ordinary Differential Equations (ODEs)

5.1 Introduction: Basic theory of ODEs

Ordinary Differential Equations (ODEs) are mathematical tools that help us understand how things change over time and used to describe dynamic processes, changes and relationships in various fields such as physics, engineering, biology, economics, science and more. They can be used to model a remarkable range of natural phenomena, such as, movement of celestial bodies, planetary system etc. But when we try to use ODEs to simulate these things on computers, it can be quite tricky because of the math involved. In this chapter, we provide a brief introduction to these issues.

An indefinite integral is the simplest representation of an ordinary differential equation. If we consider a function u(x) defined as the integral of $f(y)$ with respect to y over a certain range, denoted as, $u(x) = \int_a^x f(y)$, its derivative *u'(x)* coincides with the function $f(x)$ across the domain where the integral is well-defined. However, differential equations become more complicated when f becomes dependent on both u and x. In these cases, a basic integral isn't enough and the methodologies employed in numerical quadrature play an Important role in devising numerical techniques for tackling these complex differential equations.

5.2 Various Method to Solve ODE

There are different ways to solve Ordinary Differential Equations (ODEs), each catering to specific types of problems. Some well-known methods are the Euler method, Backward Euler method, Taylor Series Method and the First-order Exponential Integrator method. These methods help us find solutions to different types of problems. The Taylor Series Method employs polynomial approximations to estimate solutions, enabling a closer look at function behavior at different points. On the other hand, Euler's Method takes small steps to approximate solutions. When dealing with initial value problems which involve finding a function based on its starting conditions, we can use linear multistep methods and Runge-Kutta methods. Runge-Kutta methods avoid higher-order derivatives of the unknown function y and at the same time, achieves higher order of accuracy.

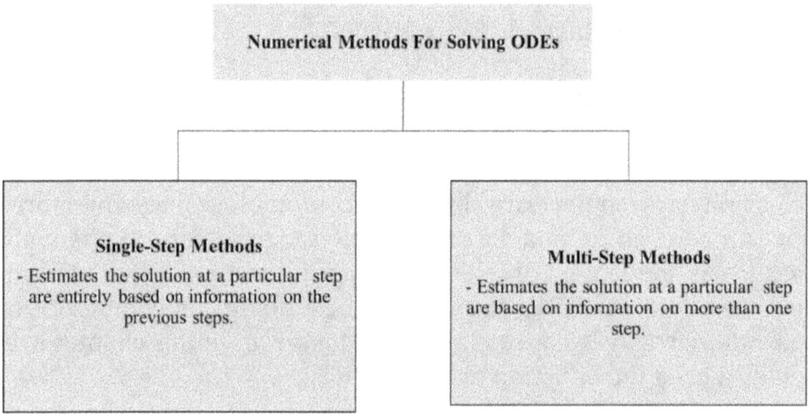

Figure 0.1 Classification of the Numerical Methods

We often use Taylor series methods and Runge-Kutta methods for starting a multistep method. Though Taylor method is conceptually easy to work with but involves higher order derivatives of the unknown function which is hard to compute and implement as computer program. An alternate solution for this is the popular Runge-Kutta methods. These methods avoid higher-order derivatives of the unknown function y and still manage to achieve higher order of accuracy. These multistep methods need a little push to begin, though—they're not self-starting. So we use a predictor method to guess what the next value of y will be and then a corrector method refines this predicted value for improved accuracy. Cubic Spline Method involves fitting a smooth curve through the data points providing a way to interpolate and approximate the behavior of the function between known points.

5.3 The Initial Value Problem

The simplest form is known as the 'initial value problem' (IVP), which involves finding a function $y(t)$. Initial value problems are a fundamental concept in engineering and are extensively used in physics. In classical mechanics, the motion of an object under the influence of a force can be determined by solving an initial value problem (IVP) using Newton's second law (F=ma), which is a second-order differential equation. Initial value problems are commonly used to represent the time evolution of a system in simplified models. Boundary value problems are particularly useful for describing equilibrium or static problems. The Diffusion equation, for instance, explains the evolution of an initial temperature distribution over time. In that case, it would be possible to specify the boundary values.

If a function $y(t)$ is such that

$$\frac{dy}{dt} = f\big(t, y(t)\big), \;\; y(t_0) = y_0 ,$$

where f is a known function from $\mathbb{R}^{N+1}$ into $\mathbb{R}^N$ for $N>0$ and N can be 1 (for a scalar equation) or greater (for a vector equation) of two variables, t is an independent variable and t_0 and y_0 are known values in $\mathbb{R}^N$. This is called the initial value problem as it starts from an initial state y_0 at an initial time t_0 (often, $t_0 = 0$). The function $y(t)$ can be a single value or a vector of values. The goal is to find the unknown function $y(t)$ that satisfies the equation,

$$y'(t) - f(t, y(t)) = 0$$

For all $t > 0$ and the initial condition $y(t_0) = y_0$. In simpler terms, we seek a function $y(t)$ that accurately describes the relationship between the rate of change of y and the values of t and y themselves.

Let's take a look at the figure below.

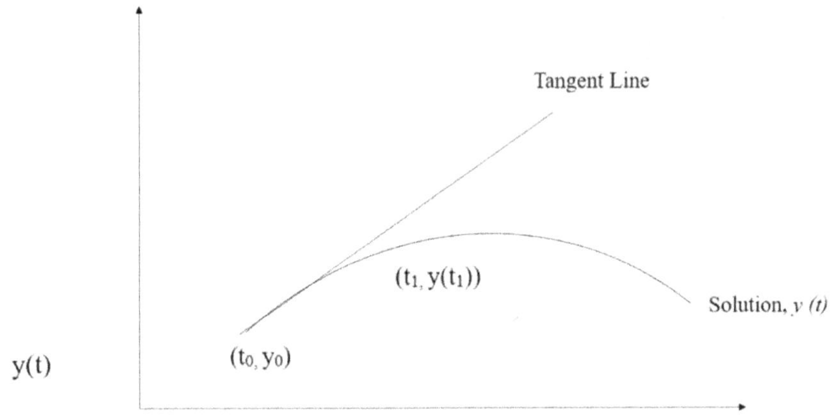

Figure 0.2 An illustration of Initial Value Problem

If t_1 is very close to t_0, then the point y_1 lying on the tangent line should be quite similar to the true value of the solution at t_1, which we call $y(t_1)$. Calculating y_1 is simple; we just need to substitute t_1 into the equation of the tangent line.

Considering y_1 is a reliable approximation, we proceed to form a line passing through the point (t_1, y_1) with a slope equal to $f(t_1, y_1)$. This leads to the equation:

$$y = y_1 + f(t_1, y_1)(t - t_1) \qquad (0.1)$$

Examples 5.1 Find the solution for the given Initial Value problem

$$y' = 3e^x + x^2 - 4, \qquad y(0) = 5$$

Solution. The first step in solving this initial-value problem is to find some general solutions. This involves finding antiderivatives of both sides of the differential equation,

$$\int y'dx = \int (3e^x + x^2 - 4)dx$$

$$y + C_1 = 3e^x + \frac{1}{3}x^3 - 4x + C_2$$

We are able to integrate both sides as the y term appears by itself. There are two integration constants: C_1 and C_2. Solving the equation for y gives,

$y = 3e^x + \frac{1}{3}x^3 - 4x + C_2 - C_1$.

As C_1 and C_2 are both constants, $C_2 - C_1$ is also a constant. We can therefore define $C = C_2 - C_1$, which leads to the equation

$$y = 3e^x + \frac{1}{3}x^3 - 4x + C$$

Now solving,

$C = 2$ and by substituting the value of C into our equation,

$$y = 3e^x + \frac{1}{3}x^3 - 4x + 2$$

Examples 5.2 Consider the Initial Value Problem

$$y' = y^{\frac{1}{3}}$$
$$y(0) = 0$$

First show that the constant solution $y(x) = 0$ satisfies this Initial Value Problem. Then see if there are any other solutions.

Solution. Certainly, if we take the derivative of a constant function we get zero, and $0^{\frac{1}{3}} = 0$, so the constant function $y(x) = 0$ satisfies the differential equation. It also satisfies the initial condition, so it does satisfy the Initial Value Problem.

This is a separable equation, so separating and integrating gives us

$$\int \frac{1}{y^{\frac{1}{3}}} dy = \int 1 dx$$

So

$$\frac{3}{2} y^{\frac{2}{3}} = x + C$$

Using the initial value $y(0) = 0$ to find C, we get

$$\frac{3}{2} 0^{\frac{2}{3}} = 0 + C,$$

So $C = 0$. That means there is a second solution to this Initial Value Problem, namely the implicit solution

$$\frac{3}{2}y^{\frac{2}{3}} = x$$

$$y^{\frac{2}{3}} = \frac{2}{3}x$$

$$y(x) = \left(\frac{2}{3}x\right)^{\frac{3}{2}}$$

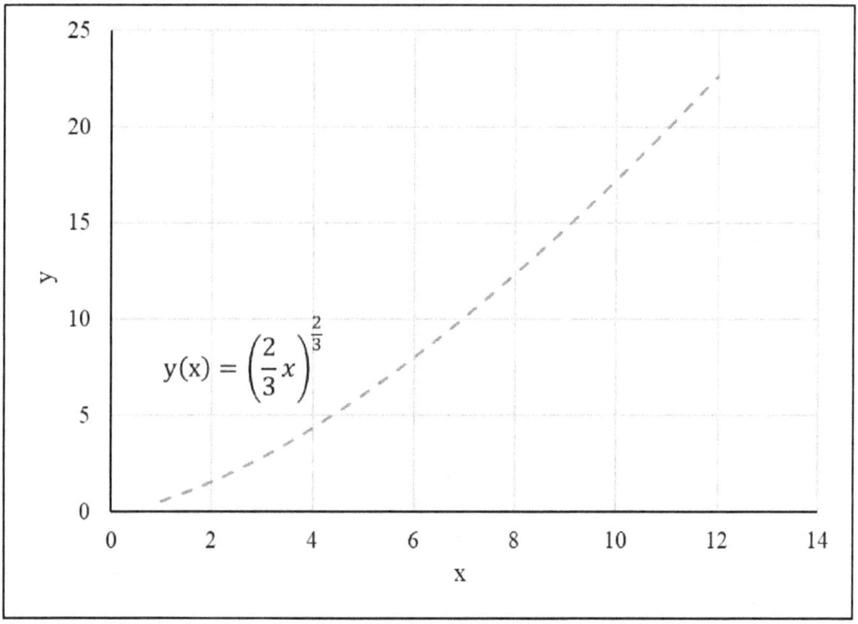

Figure 0.3 Typical graph of Initial Value Problem 5.2

It is noticed that the maximal interval of existence is $(0, \infty)$ since we can't take the square root of a negative number. If we graph the two functions, we can see that both the constant function and the function $y(x)$ that we solved for both satisfy the equation and the initial value.

Examples 5.3 Find the solution to the Initial Value Problem

$$2xy' + 4y = 3, \qquad y(2) = \frac{5}{4}$$

by first showing that the general solution is

$$y(x) = \frac{C}{x^2} + \frac{3}{4}$$

What is the interval of existence for the solution to the Initial Value Problem?

Solution. To see that $y(x)$ is a general solution, it is needed to find the derivative and plug it into the equation. The derivative is

$$y' = \frac{-2C}{x^3}$$

Plugging that in, we get

$$\begin{aligned} 2xy' + 4y &= 2x\left(\frac{-2C}{x^3}\right) + 4\left(\frac{C}{x^2} + \frac{3}{4}\right) \\ &= \frac{-4C}{x^2} + \frac{4c}{x^2} + 4\left(\frac{3}{4}\right) = 3 \ , \end{aligned}$$

which is the right-hand side of the original differential equation. Therefore $y(x)$ is a general solution.

It is noticed that the general solution is not defined at $x = 0$. Therefore, the interval of existence is either going to be $(-\infty, 0)$ or $(0, \infty)$. Since the initial value is when $x = 2$, the interval of existence for the Initial Value Problem is $(0, \infty)$.

Now using the initial value,

$$y(2) = \frac{C}{x^2} + \frac{3}{4} = \frac{5}{4}$$

Solving for C , we get, $C = 2$.

So, the solution to the Initial Value Problem is

$$y(x) = \frac{2}{x^2} + \frac{3}{4}$$

Examples 5.4 Is $y^2 = x^2 - 3$ is an implicit solution to the Initial Value Problem

$$y' = \frac{x}{y}, \qquad\qquad y(2) = 1 \ ?$$

Solution. To solve this problem, we need to apply implicit differentiation. Before proceeding, it is important to ensure that the proposed solution meets the initial value. If it does not, then it cannot be considered a solution to the initial value problem (IVP). When you check the left-hand side, you will find $\left(y(2)\right)^2 = 1^2 = 1$, and when you check the right hand side, you will find $2^2 - 3 = 1$. The proposed implicit solution, since it is the same, at least satisfies the initial value. Then using implicit differentiation and remembering that y is a function of x ,

$$\begin{aligned}\frac{d}{dx}\left(y(x)\right)^2 &= 2y(x)y'(x)\\ &= 2y\frac{x}{y}\\ &= 2x\end{aligned}$$

On the right-hand side,

$$\frac{d}{dx}(x^2 - 3) = 2x$$

Since both sides are the same, $y(x)$ does satisfy the differential equation.

Therefore, the proposed implicit solution does actually solve the Initial Value Problem.

5.4 Taylor Series Method

The Taylor series is a mathematical representation involving an infinite sum of terms. In this series, each term possesses a progressively greater exponent or degree compared to the term that came before it. This sequential progression of terms allows us to construct a polynomial approximation that converges toward the original function.

Consider the differential equation,

$$y' = f(x, y) \tag{0.2}$$

with the initial condition $y(x_0) = y_0$

If $y(x)$ is the exact solution of Eq. (5.2), then the Taylor's series for $y(x)$ around $x = x_0$ can be given by

$$y(x) = y_0 + (x - x_0)y_0' + \frac{(x-x_0)^2}{2!} y_0'' + \cdots \tag{5.3}$$

When the values of y_0', y_0'',... are known, then Eq. (5.3) gives a power series for y. By using the formula for total derivatives, we can write

$$y'' = f' = f_x + y' f_y = f_x + f f_y \tag{5.4}$$

where the suffixes denote partial derivatives with respect to the variable concerned. Similarly, it is obtained

$$\begin{aligned} y''' = f' &= f_{xx} + f_{xy} f + f\left(f_{yx} + f_{yy} f\right) + f_y\left(f_x + f_y f\right) \\ &= f_{xx} + 2 f f_{xy} + f^2 f_{yy} + f_x f_y + f {f_y}^2 \end{aligned}$$

and other higher derivatives of y. The method can also be extended to simultaneous and higher-order differential equations.

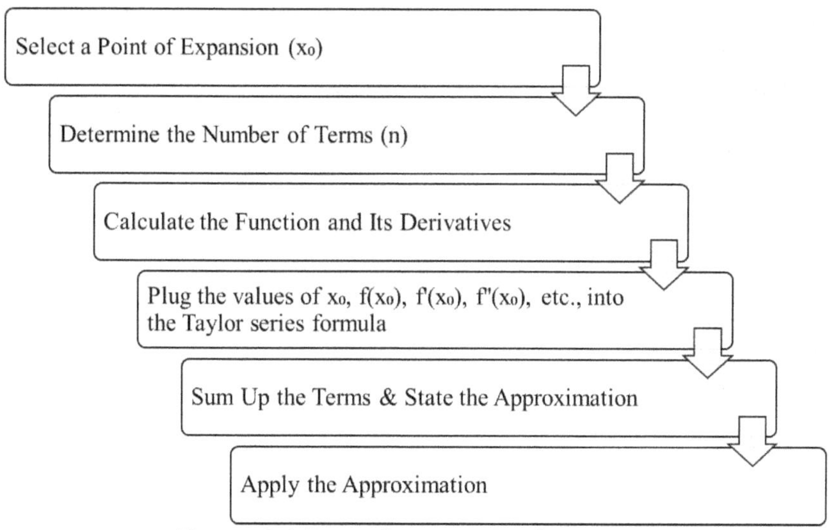

Figure 0.4 Steps for using Taylor series

Initially we need to thoroughly understand the given problem and identify the sought-after solution. Let's choose a point, denoted as x_0, around which we will expand the Taylor series. Let's formulate the Taylor series by utilizing the general expansion formula. We need to determine the appropriate number of terms (n) to include, considering the trade-off between accuracy and complexity. We will calculate the function and its derivatives at the chosen expansion point. To obtain the approximation, we need to evaluate each term in the Taylor series equation and then sum them up. Let's apply this approximation to estimate values near x_0 and assess its accuracy compared to the actual function. We can adjust the number of terms or the expansion point for improved precision if needed. Finally, we should present our solution clearly and, if possible, verify its accuracy through testing or comparisons with other methods or known results. The key to success lies in our ability to choose an appropriate expansion point and determine the number of terms that strike the right balance between simplicity and accuracy for the specific problem we are facing.

One of the most significant applications of Taylor series is their use in approximating functions by utilizing their partial sums. The partial sums mentioned are finite polynomials that can be easily calculated. These polynomials are commonly referred to as Taylor polynomials. You might be familiar with the first Taylor polynomial mentioned earlier. The study of Taylor Series is valuable because polynomial functions are relatively simple. If we can find a way to represent complex functions as infinite polynomials, known as series, it becomes much easier to analyze and understand the properties of these challenging functions.

Examples 5.5 Evaluate the Taylor Series for $f(x) = cos(x)$ for $x = 0$.

Solution. We need to take the derivatives of the cos x and evaluate them at $x = 0$.
$f(x) = cosx \Rightarrow f(0) = 1$
$f'(x) = -sinx \Rightarrow f'(0) = 0$
$f''(x) = - cosx \Rightarrow f''(0) = -1$
$f'''(x) = sinx \Rightarrow f'''(0) = 0$
$f''''(x) = cosx \Rightarrow f'''(4) = 1$
f(5)(x) = - $sinx \Rightarrow f(5)(0) = 0$
f(6) (x) = - $cosx \Rightarrow f(6)(0) = -1$

According to the Taylor series expansion;

$$cosx = \sum_{n=0}^{\infty} \frac{f^{(n)}(0)}{n!} x^n$$

$$= f(0) + f'(0)x + \frac{f''(0)}{2!}x^2 + \frac{f'''(0)}{3!}x^3 + \frac{f''''(0)}{4!}x^4 + \frac{f^5(0)}{5!}x^5 + \cdots$$

$$\cos x = 1 + \frac{x}{1!} + \frac{x^2}{2!} + \frac{x^3}{3!} + \cdots, \qquad -\infty < x < \infty$$

$$cosx = \sum_{n=0}^{\infty} \frac{(-1)^{(n)} x^{2n}}{2n!}$$

Examples 5.6 From the Taylor series for $y(x)$, find $y(0.1)$ and correct it to four decimal places if $y(x)$ satisfies $y' = x - y^2$ and $y(\mathbf{0}) = \mathbf{1}$.

Solution. The Taylor series for $y(x)$ is given by

$$y(x) = 1 + \frac{xy_0'}{1!} + \frac{x^2 y_0''}{2!} + \frac{x^3 y_0'''}{3!} + \frac{x^4 y_0''''}{4!} + \cdots$$

The derivatives $y_0', y_0'', \ldots$ etc. are obtained like this:

$$\begin{aligned} y'(x) &= x - y^2 & \qquad y_0' &= -1 \\ y''(x) &= 1 - 2yy' & y_0'' &= 3 \\ y'''(x) &= -2yy'' - 2y'^2 & y_0''' &= -8 \\ y^4(x) &= -2yy''' - 6y'y'' & y_0^4 &= 34 \\ y^5(x) &= -2yy^4 - 8y'y''' - 6y''^2 & y_0^5 &= -186 \end{aligned}$$

After using these values, the Taylor series becomes

$$y(x) = 1 - x + \frac{3x^2}{2} + \frac{4x^3}{3} + \frac{17x^4}{12} - \frac{31x^5}{20} + \cdots$$

To correct the value of $y(0.1)$ to four decimal places, it is found that the terms up to x^4 should be considered and we have $y(0.1) = 0.9138$.

Suppose that we wish to figure out the range of values of x for which the above series, truncated after the term containing x^4, can be used to compute the values of y correct to four decimal places. We just need to write

$$\frac{31x^5}{20} \leq 0.00005$$

$$\text{or,} \quad x \leq 0.126$$

Examples 5.7 Solve $y' = x + y, \qquad y(0) = 1$
by Taylor's series method. Hence find the values of y at $x = 0.1$ and $x = 0.2$.

Solution. Differentiating successively, we get

$$\begin{aligned} y' &= x + y & y'0) &= 1 \\ y'' &= 1 + y' & y''(0) &= 2 \\ y''' &= y'' & y'''(0) &= 2 \\ y''' &= y''' & y'''(0) &= 2 \end{aligned}$$

Taylor's series is

$$y = y_0 + (x - x_0)y_0' + \frac{(x - x_0)^2}{2!} y_0'' + \cdots$$

Here $x_0 = 0 \, , y_0 = 1$

$$y = 1 + x + \frac{2x^2}{2} + \frac{2x^3}{3!} + \frac{4x^4}{4!} + \cdots$$

$$y(0.1) = 1 + 0.1 + \frac{2(0.1)^2}{2} + \frac{(0.1)^3}{3!} + \frac{(0.1)^4}{4!} + \cdots$$

$= 1.1103$

And

$$y(0.2) = 1 + 0.2 + (0.2)^2 + \frac{(0.2)^3}{3} + \frac{(0.2)^4}{6} + \cdots$$

$$= 1.2427$$

5.5 Euler's Method

Euler's Method is a fundamental numerical technique for approximating solutions to Ordinary Differential Equations (ODEs) which provides a straightforward approach to understanding how functions change over time by taking small steps to model their evolution. This method is particularly useful when analytical solutions are complex or unavailable. The implementation of Euler's Method involves dividing the interval into discrete steps and at each step, updating the value of function based on its derivative.

We use Euler's Method to approximate solutions to a couple of first order differential equations. The initial equation is considered, and a decision is made regarding the use of a uniform step size. Starting from the initial point (t_0, y_0), the equations are successively computed, depending on the chosen step size method. This iterative procedure continues until the desired number of steps has been taken or the specified time is reached, resulting in a sequence of values, $y_1, y_2, y_3, \ldots, y_n$ that provide approximations of the solution at the corresponding time points $t_1, t_2, t_3, \ldots, t_n$.

Consider a differential equation,

$$y'(x) \approx \frac{1}{h}(y(x+h) - y(x))$$

Applying this approximation $x = x_j$, we get

$$\frac{1}{h}(y(x_{j+1}) - y(x_j)) \approx f(x_j, y(x_j))$$

The Euler's method is defined as follows,

$$y_{j+1} = y_j + hf(x_j, y_j),\ j = 0, 1, \cdots \tag{5.5}$$

For the initial assumption, use $y_0 = y(x_0)$.

Geometric understanding of Euler's method is provided in the following diagram.

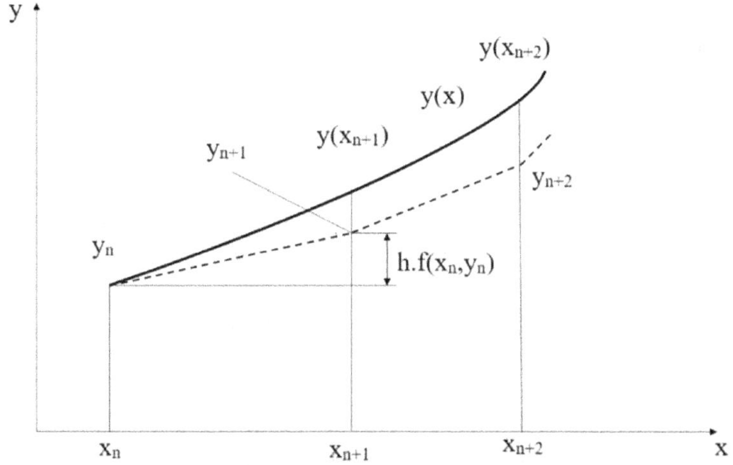

Figure 0.5 An illustration of Euler's method derivation

The slope of the tangent line to the graph of $z = y(x)$ at $x = x_n$ is $f(x_n, y_n)$. Euler's method estimates the value of $y(x_{n+1})$ by using the tangent line's value at $x = x_n + 1$.

Example 5.8 For the Initial Value Problem

$$y' + 2y = 2 - e^{-4t} \qquad y(0) = 1$$

Use Euler's Method with a step size of $h = 0.1$ to find approximate values of the solution at $t = 0.1, 0.2, 0.3, 0.4$, and 0.5. Compare them to the exact values of the solution at these points.

Solution. This is a fairly simple linear differential equation so we'll leave it to you to check that the solution is

$$y(t) = 1 + \frac{1}{2}e^{-4t} - \frac{1}{2}e^{-2t}$$

In order to use Euler's Method, we first need to rewrite the given differential equation,

$$y' = 2 - e^{-4t} - 2y$$

From this we can see that,

$$f(t, y) = 2 - e^{-4t} - 2y$$

For $t_0 = 0$ and $y_0 = 0$, we can write,

$$f_0 = f(0,1) = 2 - e^{-4(0)} - 2(1) = -1$$

So, the approximation to the solution at

$y_1 = y_0 + hf_0 = 1 + (0.1)(-1) = 0.9$

So, the approximation to the solution at $t_1 = 0.1$ and $y_1 = 0.9$

At the next step we have,

$$f_1 = f(0.1,0.9) = 2 - e^{-4(0.1)} - 2(0.9) = -0.470320046$$

$$y_2 = y_1 + hf_1 = 0.9 + (0.1)(-0.470320046) = 0.852967995$$

Therefore, the approximated solution at $t_2 = 0.2$ is $y_2 = 0.852967995$

The approximations as well as the exact value of the solutions at the given points are presented in the following table.

Table 0.1 Approximations for problem 5.8

Time, t_n	Approximation	Exact	Error
$t_0 = 0$	$y_0 = 1$	$y(t) = 0$	0%
$t_1 = 0.1$	$y_1 = 0.9$	$y(0.1) = 0.925794646$	2.79 %
$t_2 = 0.2$	$y_2 = 0.852967995$	y(0.2)=0.889504459	4.11 %
$t_3 = 0.3$	$y_3 = 0.837441500$	$y(0.3) = 0.876191288$	4.42 %
$t_4 = 0.4$	$y_4 = 0.839833779$	$y(0.4) = 0.876283777$	4.16 %
$t_5 = 0.5$	$y_5 = 00.851677371$	$y(0.5) = 0.883727921$	3.63 %

The formula for the error as a percentage is,

$$percent\ error = \frac{|exact - approximate|}{exact} \times 100$$

From the above numerical experiments, we could see that to obtain more accuracy with Euler's method, we must approximate a considerably smaller value.

Example 5.9 Repeat the previous example with the given approximations at $t = 1$, $t = 2$, $t = 3$, $t = 4$, and $t = 5$. Use $h = 0.1$, $h = 0.05$, $h = 0.01$, $h = 0.005$, and $h = 0.001$ for the approximations.

Solution. Here are two tables: one showing the approximations for the solution, and the other indicating the associated errors. The task of verifying the computational intricacies is deferred for your consideration.

Table 0.2 Approximations for problem 5.9

Time	Exact	$h = 0.1$	$h = 0.05$	$h = 0.01$	$h = 0.005$	$h = 0.001$
t =1	0.9414902	0.9313244	0.9364698	0.9404994	0.9409957	0.9413914
$t = 2$	0.9910099	0.9913681	0.9911126	0.9910193	0.9910139	0.9910106
$t = 3$	0.9987637	0.9990501	0.9988982	0.9987890	0.9987763	0.9987662
$t = 4$	0.9998323	0.9998976	0.9998657	0.9998390	0.9998357	0.9998330
$t = 5$	0.9999773	0.9999890	0.9999837	0.9999786	0.9999780	0.9999774

Table 0.3 Percentage error for each approximation

Time	$h = 0.1$	$h = 0.05$	$h = 0.01$	$h = 0.005$	$h = 0.001$
t =1	1.08 %	0.53 %	0.105 %	0.053 %	0.0105 %
$t = 2$	0.036 %	0.010 %	0.00094 %	0.00041 %	0.0000703 %
$t = 3$	0.029 %	0.013 %	0.0025 %	0.0013 %	0.00025 %
$t = 4$	0.0065 %	0.0033 %	0.00067 %	0.00034 %	0.000067 %
$t = 5$	0.0012 %	0.00064 %	0.00013 %	0.000068 %	0.000014 %

It can be seen from these tables that decreasing h does in fact improve the accuracy of the approximation as we expected.

Example 5.10 For the Initial Value Problem

$$y' - y = -\frac{1}{2}e^{\frac{t}{2}}sin(5t) + 5e^{\frac{t}{2}}cos(5t) \qquad y(0) = 0$$

Use Euler's Method to find the approximation to the solution at t =1, $t = 2$, $t = 3$, $t = 4$, and $t = 5$. Use $h = 0.1$, $h = 0.05$, $h = 0.01$, $h = 0.005$, and $h = 0.001$ for the approximations.

Solution. The solution to this linear first order differential equation is

$$y(t) = e^{\frac{t}{2}}sin(5t)$$

Two tables are given below showing the approximations and the percentage error for each approximation.

Table 0.4 Approximations for problem 5.10

Time	Exact	$h = 0.1$	$h = 0.05$	$h = 0.01$	$h = 0.005$	$h = 0.001$
t =1	-1.58100	-0.97167	-1.26512	-1.51580	-1.54826	-1.57443
$t = 2$	-1.47880	0.65270	-0.34327	-1.23907	-1.35810	-1.45453
$t = 3$	2.91439	7.30209	5.34682	3.44488	3.18259	2.96851
$t = 4$	6.74580	15.56128	11.84839	7.89808	7.33093	6.86429
$t = 5$	-1.61237	21.95465	12.24018	1.56056	0.0018864	-1.28498

Table 0.5 Percentage error for each approximation

Time	$h = 0.1$	$h = 0.05$	$h = 0.01$	$h = 0.005$	$h = 0.001$
$t = 1$	38.54 %	19.98 %	4.12 %	2.07 %	0.42 %
$t = 2$	144.14 %	76.79 %	16.21 %	8.16 %	1.64 %
$t = 3$	150.55 %	83.46 %	18.20 %	9.20 %	1.86 %
$t = 4$	130.68 %	75.64 %	17.08 %	8.67 %	1.76 %
$t = 5$	1461.63 %	859.14 %	196.79 %	100.12 %	20.30 %

Therefore, Euler's Method does not perform nearly as well as it did on the Initial Value Problem. However, some of the observations we made in Example 5.5.3 still hold true. Reducing the magnitude of h reduces the error, as observed in the previous example and as would be expected. Also, as we saw in the previous example, decreasing h by a factor of 10 reduces the error by roughly the same factor.

Unlike the previous example, however, increasing t results in an increasing error. This is a reasonably common occurrence in the approximations. We should not anticipate that the error will decrease as t increases, as we saw in the previous example. Each successive approximation is derived from the previous one. As a result, we introduce error at each stage, and approximations should, in general, become less accurate as t increases.

5.6 Runge-Kutta Method

The Runge-Kutta methods are widely recognized as one of the most commonly used techniques for solving ordinary differential equations (ODEs). The initial investigation of these entities was conducted by Carle Runge and Martin Kutta in 1900. The predominant advancements in contemporary times can be primarily attributed to the contributions made by John Butcher during the 1960s.

The application of the Taylor series approach for numerical solution of differential equations is limited due to the computational effort required to determine the higher order derivatives. The Runge–Kutta method is an efficient and widely-applied technique for solving initial-value problems involving differential equations. Runge–Kutta method can be used to build high order accurate numerical methods without the need for high order derivatives of functions.

However, there exists a category of techniques referred to as Runge-Kutta methods that obviate the need for computing higher order derivatives while simultaneously providing enhanced precision. The Runge-Kutta formulas offer the benefit of solely necessitating the function values at specific places. These approaches are consistent with Taylor's series solution, specifically up to the term in h^r. It is important to note that the value of r varies depending on the specific method being utilized, and this value is referred to as the order of the method.

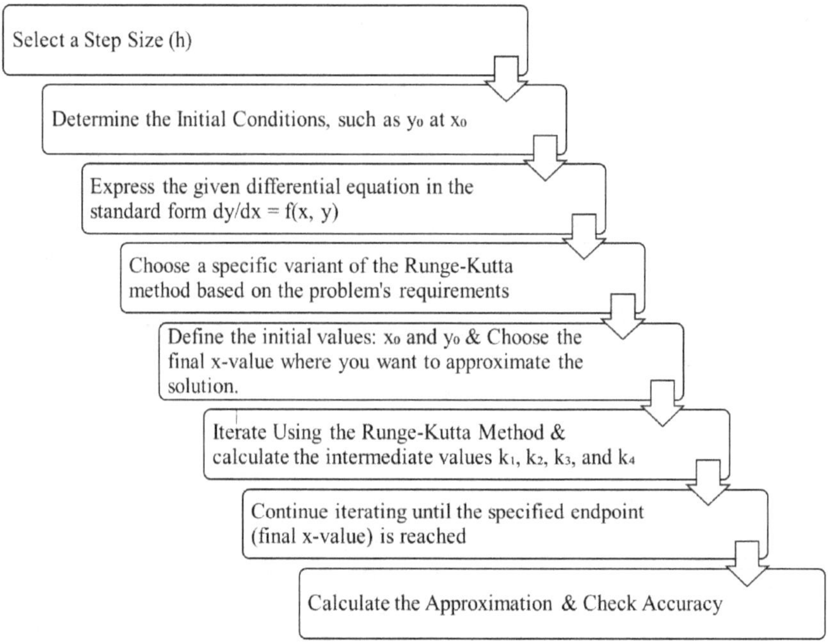

Figure 0.6 Steps for using Runge-Kutta method

Solving mathematical problems using the Runge-Kutta method requires a systematic approach. First, we need to understand the problem statement and determine the specific initial conditions, including the starting point (x_0) and the corresponding function value (y_0). Let's express the given differential equation in the form $\frac{dy}{dx} = f(x, y)$, which represents the rate of change of y with respect to x. Next, we will choose an appropriate Runge-Kutta method, such as the popular RK4, tailored to our problem's requirements. With these preparations in place, let's set up the iteration process, specifying the initial values and the final x-value we aim to approximate. The heart of our method lies in the iterative loop, where we continually update the values of x and y using intermediate calculations until we reach the defined endpoint. The resulting x and y values provide us with a numerical approximation of the solution. After calculating the approximation, we must assess its accuracy, which hinges on factors such as step size and method choice. Finally, we should present our solution clearly, potentially comparing it with known results or using validation tests to ensure its correctness.

5.6.1 First Order Runge-Kutta Method

We have seen that Euler's method gives

$$y_1 = y_0 + hf(x_0, y_0) = y_0 + hy_0' \qquad [y' = f(x,y)]$$

Expanding it by Taylor's series,

$$y_1 = y(x_0 + h) = y_0 + hy_0' + \frac{h^2}{2!}y_0'' + \cdots \tag{5.6}$$

The Euler method is consistent with Taylor's series solution up to the term in h. Therefore, it can be stated that Euler's approach is equivalent to the first-order Runge-Kutta method.

5.6.2 Second Order Runge-Kutta Method

The modified Euler's method gives

$$y_1 = y + \frac{h}{2}[f(x_0, y_0) + f(x_0 + h, y_1)] \tag{5.7}$$

Substituting $y_1 = y_0 + hf(x_0, y_0)$ on the right-hand side of (5.7), we obtain

$$y_1 = y_0 + \frac{h}{2}[f_0 + f(x_0 + h), y_0 + hf_0)] \qquad \text{where} \qquad f = (x_0, y_0)$$
(5.8)

Expanding L.H.S. by Taylor's series, we get

$$y_1 = y(x_0 + h) = y_0 + hy_0' + \frac{h^2}{2!}y_0'' + \frac{h^3}{3!}y_0''' + \cdots \tag{5.9}$$

Expanding $f(x_0 + h), y_0 + hf_0)$ by Taylor's series for a function of two variables, (5.8) gives

$$y_1 = y_0 + \frac{h}{2}\left[f_0 + \left\{f_0 = (x_0, y_0) + h\left(\frac{\partial f}{\partial x}\right)_0 + hf_0\left(\frac{\partial f}{\partial y}\right)_0 + O(h^2)\right\}\right]$$

$$= y_0 + \frac{1}{2}\left[hf_0 + hf_0 + h^2\left\{\left(\frac{\partial f}{\partial x}\right)_0 + \left(\frac{\partial f}{\partial y}\right)_0\right\} + O(h^3)\right]$$

$$= y_0 + hf_0 + \frac{h^2}{2}f_0'$$

$$+ O(h^3) \qquad \left[\because \frac{df(x,y)}{dx} = \frac{\partial f}{\partial x} + f\frac{\partial f}{\partial y}\right]$$

$$= y_0 + hy_0' + \frac{h^2}{2!}y_0'' + O(h^3) \tag{5.10}$$

Comparing equations (5.9) and (5.10), it follows that the modified Euler's method accords with Taylor's series solution up to the term in h^2.

Therefore, it can be stated that the modified Euler's approach is equivalent to the second-order Runge-Kutta method.

$\therefore$ The second order Runge-Kutta formula is

$$y_1 = y_0 + \frac{1}{2}(k_1 + k_2) \tag{5.11}$$

Where $k_1 = hf(x_0, y_0)$ and $k_2 = hf(x_0 + h, y_0 + k)$

5.6.3 Third Order Runge-Kutta Method

Similarly, Runge's method is consistent with Taylor's series solution up to the term in h^3. Therefore, Runge's method is the third-order Runge-Kutta method.

$\therefore$ The third order Runge-Kutta formula is

$$y_1 = y_0 + \frac{1}{6}(k_1 + 4k_2 + k_3) \tag{5.12}$$

Where $k_1 = hf(x_0, y_0)$, $k_2 = hf\left(x_0 + \frac{1}{2}h, y_0 + \frac{1}{2}k_1\right)$
And $k_3 = hf(x_0 + h, y_0 + k')$ where $k' = k_3 = hf(x_0 + h, y_0 + k_1)$.

5.6.4 Fourth Order Runge-Kutta Method

This is the most often used method, and it is typically referred to only as the Runge-Kutta method.

Working rule for using the Runge-Kutta method to find the increment k of y that corresponds to an increase h of x from

$$\frac{dy}{dx} = f(x, y), y(x_0)$$

is as follows:

$$k_1 = hf(x_0, y_0)$$
$$k_2 = hf\left(x_0 + \frac{1}{2}h, y_0 + \frac{1}{2}k_1\right)$$
$$k_3 = hf\left(x_0 + \frac{1}{2}h, y_0 + \frac{1}{2}k_2\right)$$
$$k_4 = hf(x_0 + h, y_0 + k_3)$$

Then, finally compute $k = \frac{1}{6}(k_1 + 2k_2 + 2k_3 + k_4)$
which gives the required approximate value as $y_1 = y_0 + k$.

Example 5.11 Apply the Runge-Kutta fourth order method to find an approximate value of y when $x = 0.2$ given that $\frac{dy}{dx} = x + y$ and $y = 1$ when $x = 0.$

Solution. Here, $x_0 = 0, h = 0.2, f(x_0, y_0) = 1$
$\therefore\ k_1 = hf(x_0, y_0) = 0.2 \times 1 = 0.2000$
$k_2 = hf\left(x_0 + \frac{1}{2}h, y_0 + \frac{1}{2}k_1\right) = 0.2 \times f(0.1, 1.1) = 0.2400$
$k_3 = hf\left(x_0 + \frac{1}{2}h, y_0 + \frac{1}{2}k_2\right) = 0.2 \times f(0.1, 1.1) = 0.2400$
And $k_4 = hf(x_0 + h, y_0 + k_3) = 0.2 \times f(0.2, 1.244) = 0.2888$
Therefore,

$$k = \frac{1}{6}(k_1 + 2k_2 + 2k_3 + k_4)$$
$$= \frac{1}{6}(0.2000 + 0.2400 + 0.2888) = 0.2428$$

Hence the required approximate value of y is 1.2428.

Example 5.12 Using the Runge-Kutta method of fourth order, solve

$$\frac{dy}{dx} = \frac{y^2 - x^2}{y^2 + x^2}$$

With $y(0) = 1$ at $x = 0.2, 0.4.$

Solution. We have

$$f(x,y) = \frac{y^2 - x^2}{y^2 + x^2}$$

To find $y(0.2)$

Hence $x_0 = 0, y_0 = 0, h = 0.2$

$k_1 = hf(x_0, y_0) = 0.2 \times \mathrm{f}(0,1) = 0.2000$

$k_2 = hf\left(x_0 + \frac{1}{2}h, y_0 + \frac{1}{2}k_1\right) = 0.2 \times f(0.1,1.1) = 0.19672$

$k_3 = hf\left(x_0 + \frac{1}{2}h, y_0 + \frac{1}{2}k_2\right) = 0.2 \times f(0.1,1.09836) = 0.1967$

$k_4 = hf(x_0 + h, y_0 + k_3) = 0.2 \times f(0.2, 0.1967) = 0.1891$

$k = \frac{1}{6}(k_1 + 2k_2 + 2k_3 + k_4)$

$= \frac{1}{6}(0.2 + 2(0.19672) + 2(0.1967) + 0.1891) = 0.19599$

Hence $y(0.2) = y_0 + k = 1.196$

To find $y(0.4)$,

Here $x_1 = 0.2, y_0 = 1.196, h = 0.2.$

$k_1 = hf(x_1, y_1) = 0.1891$

$k_2 = hf\left(x_1 + \frac{1}{2}h, y_1 + \frac{1}{2}k_1\right) = 0.2 \times f(0.3,1.2906) = 0.1795$

$k_3 = hf\left(x_1 + \frac{1}{2}h, y_1 + \frac{1}{2}k_2\right) = 0.2 \times f(0.3,1.25858) = 0.1793$

$k_4 = hf(x_1 + h, y_1 + k_3) = 0.2 \times f(0.4, 1.3753) = 0.1688$

$k = \frac{1}{6}(k_1 + 2k_2 + 2k_3 + k_4)$

$= \frac{1}{6}(0.1891 + 2(0.1795) + 2(0.1793) + 0.1688) = 0.1792$

Hence $y(0.4) = y_1 + k = 1.196 + 0.1792 = 1.3752$

Example 5.13 Apply the Runge-Kutta method to find the approximate value of y for $x = 0.2$, in steps of 0.1, if $\frac{dy}{dx} = x + y^2, y = 1$ where $x = 0$.

Solution: Given, $f(x,y) = x + y^2$.

Here we assume that $h = 0.1$ and carry out the calculations in two steps.

Step I. $\quad x_0 = 0, y_0 = 0, h = 0.1$

$k_1 = hf(x_0, y_0) = 0.1 \times \text{f}(0,1) = 0.1000$

$k_2 = hf\left(x_0 + \frac{1}{2}h, y_0 + \frac{1}{2}k_1\right) = 0.1 \times f(0.05,1.1) = 0.1152$

$k_3 = hf\left(x_0 + \frac{1}{2}h, y_0 + \frac{1}{2}k_2\right) = 0.1 \times f(0.05,1.1152) = 0.1168$

$k_4 = hf(x_0 + h, y_0 + k_3) = 0.1 \times f(0.1, 1.1168) = 0.1347$

$k = \frac{1}{6}(k_1 + 2k_2 + 2k_3 + k_4)$

$= \frac{1}{6}(0.1000 + 0.2304 + 0.2336 + 0.1347) = 0.1165$

Giving $y(0.1) = y_0 + k = 1.1165$

Step II. $\quad x_1 = 0.1\,, y_1 = 1.1165,\ h = 0.1$

$k_1 = hf(x_1, y_1) = 0.1347$

$k_2 = hf\left(x_1 + \frac{1}{2}h, y_1 + \frac{1}{2}k_1\right) = 0.1 \times f(0.15,1.1838) = 0.1551$

$k_3 = hf\left(x_1 + \frac{1}{2}h, y_1 + \frac{1}{2}k_2\right) = 0.1 \times f(0.15,1.194) = 0.1576$

$k_4 = hf(x_1 + h, y_1 + k_3) = 0.1 \times f(0.2, 1.1576) = 0.1823$

$k = \frac{1}{6}(k_1 + 2k_2 + 2k_3 + k_4)$

$= 0.1571$

Hence $y(0.2) = y_1 + k = 1.1165 + 0.1571 = 1.2736$

Example 5.14 Using the Runge-Kutta method of fourth order, solve for y at $x = 1.2, 1.4$. From

$$\frac{dy}{dx} = \frac{2xy + e^x}{x^2 + xe^x}$$

Given, $x_0 = 1,\ \ y_0 = 0$

Solution. We have

$$f(x,y) = \frac{2xy + e^x}{x^2 + xe^x}$$

To find $y(1.2)$,

Here, $x_0 = 1,\ \ y_0 = 0, h = 0.2$

$$k_1 = hf(x_0, y_0) = 0.2\frac{0 + e'}{1 + e'} = 0.1462$$
$$k_2 = hf\left(x_0 + \frac{1}{2}h, y_0 + \frac{1}{2}k_1\right) = 0.1402$$
$$k_3 = hf\left(x_0 + \frac{1}{2}h, y_0 + \frac{1}{2}k_2\right) = 0.1399$$
$$k_4 = hf(x_0 + h, y_0 + k_3) = 0.1348$$
$$k = \frac{1}{6}(k_1 + 2k_2 + 2k_3 + k_4) = 0.1402$$

Hence, $y(1.2) = y_0 + k = 0 + 0.1402$.

To find $y(1.4)$,

Here $x_1 = 1.2$, $y_1 = 0.1402$, $h = 0.2$
$$k_1 = hf(x_1, y_1) = 0.2f(1.2,0) = 0.1348$$
$$k_2 = hf\left(x_1 + \frac{1}{2}h, y_1 + \frac{1}{2}k_1\right) = 0.2 \times f(1.3,0.2076) = 0.1303$$
$$k_3 = hf\left(x_1 + \frac{1}{2}h, y_1 + \frac{1}{2}k_2\right) = 0.2 \times f(1.3,0.2053) = 0.1301$$
$$k_4 = hf(x_1 + h, y_1 + k_3) = 0.2 \times f(1.3, 0.2703) = 0.1260$$
$$k = \frac{1}{6}(k_1 + 2k_2 + 2k_3 + k_4)$$
$$= 0.1303$$

Hence, $y(1.4) = y_1 + k = 0.1402 + 0.1303 = 0.2705$.

5.7 Multistep Methods: Predictor and Corrector

Predictor-corrector methods are a type of algorithm used in numerical analysis. These algorithms are specifically designed to solve ordinary differential equations by finding an unknown function that satisfies a given differential equation. All of these algorithms follow a two-step process. The first step, known as the "prediction" step, involves using a function that has been fitted to the function-values and derivative-values at a previous set of points. This allows us to estimate or anticipate the value of this function at a new point in the future. The next step, known as the "corrector" step, improves the initial approximation by utilizing the predicted value of the function and another interpolation method to estimate the value of the unknown function at the same subsequent point.

Multi-step methods differ from single-step methods in that they utilize multiple previous steps to calculate the next value of y. The multi-step method can be expressed in a general form

$$y_{n+m} + a_{m-1}y_{n+m-1} + \cdots + a_0 y_n = h\big(b_m f(x_{n+m}, y_{n+m}) + b_{m-1} f(x_{n+m-1}, y_{n+m-1}) + \cdots + b_0 f(x_n, y_n)\big) \tag{5.13}$$

Where a_j, $(j = 0,1, \ldots, m-1)$ and $b_i, (i = o, 1, \ldots, m)$ are constants to be determined.

$m = 1, a_0 = -1\,, b_0 = 1$ and $b_1 = 0$ gives Euler method.

Similarly, $m = 1, a_0 = -1\,, b_0 = \frac{1}{2}$ and $b_1 = \frac{1}{2}$ gives the trapezoidal method. When $b_m = 0$, the method is called the **Explicit method** and when $b_m \neq 0$, the method is called the implicit method. The explicit methods are generally called as the Adams-Bashforth methods and the implicit methods are called the **Adams-Moulton methods**.

Implicit multi-step methods are commonly used to enhance the accuracy of approximations obtained through explicit methods. The procedure involves two steps. First, an approximation y_{n+m} for the solution at x_{n+m} is obtained using an explicit method known as the predictor step. Then, this value is iteratively used in the implicit method to improve the approximation, which is referred to as the corrector step. The method that is obtained is known as the predictor-corrector method.

Let's start by explaining the concept of a single step predictor-corrector method. In the predictor step, we utilize the Euler method

$$y_{n+1} = y_n + hf(x_n\,, y_n), \qquad n = 0, 1, \cdots$$

which approximates the solution $y(x_{n+1})$ to the provided initial value problem.

Consider this value to be the initial term of the sequence $\{y_{n+1}^{(k)}$, where $k = 0,1, \ldots$ and generate the remaining terms using the implicit trapezoidal rule as follows:

$$y_{n+1}^{(k+1)} = y_n + \frac{h}{2}\left(f(x_n, y_n) + f\left(x_{n+1}, y_{n+1}^{(k)}\right)\right), \qquad k = 0,1,\dots$$

(5.14)

For the multi-step predictor-corrector method, the coefficients b_m on the right side of the equation, must be determined. To obtain, we integrate the equation $y' = f(x, y)$ over the interval $[x_n, x_{n+1}]$, where $x_{n+1} = x_n + h$.

$$y(x_{n+1}) - f(x_n) = \int_{x_n}^{x_{n+1}} f(x, y(x))dx \qquad (5.15)$$

The idea of a general mutistep method is to replace $f(x, y(x))$ by an interpolation polynomial $p(x)$ and then integrate. This gives approximations y_{n+1} of $y(x_{n+1})$ and y_n of $y(x_n)$.

$$y_{n+1} = y_n + \int_{x_n}^{x_{n+1}} p(x)dx \qquad (5.16)$$

Different selections of $p(x)$ will result in distinct approaches.

Let's proceed with the derivation of the three-step predictor-corrector method. Let $p_2(x)$ represent the quadratic polynomial that interpolates $f(x, y(x)) = g(x)$ at x_n, x_{n-1} and x_{n-2}. Then use,

$$\int_{x_n}^{x_{n+1}} g(x)dx \approx \int_{x_n}^{x_{n+1}} p_2(x)dx.$$

The final expression together with truncation error is

$$\int_{x_n}^{x_{n+1}} g(x)dx = \frac{h}{12}[23g(x_n) - 16g(x_{n-1}) + 5g(x_{n-2})] + \frac{3}{8}h^4 g'''(\xi_n), \qquad (5.17)$$

Where $x_n < \xi_n < x_{n+1}$.

For the predictor step (Adams-Bashforth method), we use,

$$y_{n+1} = y_n + \frac{h}{12}[23f_n - 16f_{n-1} + 5f_{n-2}], \quad n \geq 2,$$

(5.18)

Where $f_n = f(x_n, y_n)$.

The corrector step (Adams-Moulton method) is acquired by replacing the quadratic interpolation $p_2(x)$ for the function $f(x, y(x)) = g(x)$ at x_{n+1}, x_n and x_{n-1}. The resulting method is,

$$y_{n+1} = y_n + \frac{h}{12}[5f_{n+1} + 8f_n - f_{n-1}], \quad n \geq 1, \tag{5.19}$$

Where $f_n = f(x_n, y_n)$.

Example 5.15 Find the approximate value of $y(1)$ using the multistep method (Adams-Bashforth method)

$$y_{i+1} = y_i + \frac{h}{2}(3f_i - f_{i-1})$$

With $h = 0.2$ for the initial value problem

$$y' = -2xy^2, y(0) = 1$$

Solution. For the given multistep method

$$y_{i+1} = y_i + \frac{h}{2}(3f_i - f_{i-1})$$

Or,

$$y_{i+2} = y_{i+1} + \frac{h}{2}(3f_{i+1} - f_i)$$

We get for $i = 0$

$$y_2 = y_1 + \frac{h}{2}(3f_1 - f_0)$$

Now y_0 is the given initial value and y_1 is the starting value. We calculate the starting value using the given second order Runge-Kutta method. We obtain from

$$f(x, y) = -2xy^2, x_0 = 0, y_0 = 1, h = 0.2$$

$$k_1 = hf(x_0, y_0) = hf(0,1) = 0$$

$$k_2 = hf\left(x_0 + \frac{1}{2}h, y_0 + \frac{1}{2}k_1\right) = hf(0.2,1) = -0.08$$

$$y(0.2) = y_0 + \frac{1}{2}(k_1 + k_2) = 0.96$$

After the starting value has been determined, we use the multistep method. We obtain for

$i = 0$: $x_0 = 0, y_0 = 1, f_0 = -2x_0{y_0}^2 = 0$

$x_1 = 0.2, y_1 = 0.96, f_1 = -2x_1{y_1}^2 = -0.36864$

$$y_2 = y(0.4) = y_1 + \frac{h}{2}(3f_1 - f_0) = 0.849408$$

$i = 1$: $x_2 = 0.4, y_2 = 0.849408, f_2 = -2x_2{y_2}^2 = -0.577195$

$$y_3 = y(0.6) = y_2 + \frac{h}{2}(3f_2 - f_1) = 0.713114$$

$i = 2$: $x_3 = 0.6, y_3 = 0.713114, f_3 = -2x_3{y_3}^2 = -0.610237$

$$y_4 = y(0.8) = y_3 + \frac{h}{2}(3f_3 - f_2) = 0.587762$$

$i = 3$: $x_4 = 0.8, y_4 = 0.587762, f_4 = -2x_4{y_4}^2 = -0.552743$

$$y_5 = y(1.0) = y_4 + \frac{h}{2}(3f_4 - f_3) = 0.482963\,.$$

5.8 Exercises

1. For each initial value problem below, verify (by direct substitution) that the given function y solves the problem.

a) $y' + 4y = 0,\ y(0) = 3;\ \ y(t) = 3e^{-4t}$;

b) $y' = \frac{t^2}{y}\ \ y(0) = 1,\ y(t) = \sqrt{1 + \frac{2}{3}t^3}$;

c) $ty' - y = t^2,\ y(1) = 4;\ \ y(t) = 3t + t^2$.

2. For each initial value problem in Problem 1, write down the definition of f .

3. Consider the initial value problem $y' = xy\,,\ y(0) = 1$. Estimate the error at $x = 1$ when Euler method is used with infinite precision, to find the approximate solution to this problem with step size $h = 0.01$.

4. Consider the initial value problem $y' = -2y, 0 \le x \le 1$, $y(0) = 1$;

a) Find an upper bound on the error in Euler method at $x = 1$ in terms of the step size h.

b) Solve the difference equation which results from Euler's method.

c) Compare the bound obtained from (a) with the actual error as obtained from (b) at $x = 1$ for $h = 0.1$ and $h = 0.01$.

d) How small a step size h would have to be taken to produce six significant digits of accuracy at $x = 1$, using Euler's method?

5. Use Euler, Range-Kutta of order 2 and 4 methods to solve the Initial Value Problem $y' = 0.5(x - y)$ for all $x \in [0, 3]$ with initial condition $y(0) = 1$. Compare the solutions for $h = 1, 0.5, 0.25, 0.125$ along with the exact solution $y(x) = 3e^{-x/2} + x - 2$.

6. Find the Taylor series expansion of $ln(1 + x)$ at $x = 2$.

7. Evaluate the Taylor series expansion for $f(x) = x^3 - 10x^2 + 6$ at $x = 3$.

8. Find $y(0.2)$ for $y' = x^2y - 1$, $y(0) = 1$, with step length 0.1 using Taylor Series method.

9. Find $y(0.5)$ for $y'' = -2x - y, y(0) = -1$, with step length 0.1 using Taylor Series method.

10. Find $y(0.2)$ for $y' = \frac{x-y}{2}$, $y(0) = 1$, with step length 0.1 using Euler method.

11. Find $y(0.5)$ for $y'' = -2x - y, y(0) = -1$, with step length 0.1 using Euler method.

12. Find $y(0.1)$ for $y'' = 1 + 2xy - x^2z$, , with step length 0.1 using Euler method (2nd order derivative).

13. In each of the following initial value problems, use Euler method, Runge-Kutta method of order 2 and 4 to find the solution at the specified point with specified step size h:
a) $y' = x + y$; $y(0) = 1$. Find $y(0.2)$. (For Euler method take $h = 0.1$ and for other methods, take $h = 0.2$) Exact Solution: $y(x) = -1 - x + 2e^x$
b) $y' = 2\cos x - y, y(0) = 1$. Find $y(0.6)$ (For Euler method take $h = 0.1$ and for other methods, take $h = 0.2$) Exact Solution: $y(x) = \sin x + \cos x$

14. Find $y(0.1)$ for $y'' = 1 + 2xy - x^2z$, $x_0 = 0, y_0 = 1, z_0 = 0$, with step length 0.1 using Runge-Kutta 3 method (2nd order derivative).

15. Given the velocity, $v = \frac{ds}{dt} = 32t - 2$,and the initial position of the body as $s\ (1/2)\ = 4$. Find the body's position at time t.

16. Solve the initial value problem $y^4 = -sint + cost, y'''(0) = 7, y''(0) = -1, y(0) = 0$.

17. Find the expansion for the function, $f(x)\ = 4x$ centred at $a\ = 1$ using the Taylor Series Formula.

18. Write the Euler method for finding the approximate value of $y(x_n)$ for some $x_n < 0$, where y satisfies the initial value problem $y(x)\ = f(x,y), y(0)\ = y_0$.

19. Write the Euler method for finding the approximate value of $y(x_n)$ for some $x_n < 0$, where y satisfies the initial value problem $y(x) = f(x,y)\ ,\ y(0) = y_0$.

20. Given that,

$$\frac{dy}{dx} = \frac{(y^2 - 2x)}{(y^2 + x)}$$

And $y = 1$ at $x = 0$, find y for $x = 0.1, 0.2, 0.3, 0.4$, and 0.5 .

21. Use Euler's method with $h = \frac{1}{4}$ to compute approximate values of $y(1)$ for each of the following initial value problems. Don't write a computer program, use a hand calculator to produce an orderly table of (t_k, y_k) pairs.

a) $y' = y(1 - y),\ y(0)\ = \frac{1}{2}$;
b) $ty' = y(sint),\ y(0) = 2$;
c) $y' = y(1 + e^{2t}),\ y(0) = 1$;
d) $y' + 2y\ = 1,\ y(0)\ = 2$.

22. For each initial value problem above, use the differential equation to produce approximate values of y' at each of the grid points, $t_k, k = 0,1,2,3,4$

23. Use Taylor's Theorem to prove that

$$(1 + x)^n \leq e^{nx}$$

24. Consider the initial value problem

$$y' = e^{-t} - 16y, \;\; y(0) = 1\,.$$

a) Confirm that this is smooth and uniformly monotone decreasing in y. What is M ? What is m ?

b) Approximate the solution using Euler's method and $h = \frac{1}{8}$. Do we get the expected behavior from the approximation? Explain.

25. Using the Runge-Kutta method of order 4, find y for $x = 0.1, 0.2, 0.3$ given that $\frac{dy}{dx} = xy + y^2$, $y(0) = 1$.

26. Use Runge's method to approximate y where $x = 0.11$, given that $y = 1.2$ When $x = 1$ and $\frac{dy}{dx} = 3x + y^2$.

27. Using the Runge-Kutta method, calculate $y\,(0.1), y(0.2)$, and $y(0.3)$ given that $\frac{dy}{dx} = \frac{2xy}{1+x^2} = 1\,,\; y(0) = 0\,.$

28. Using Taylor's series method, solve

$dy/_{dx} = xy + x^2$, $y(0) = 1$; at $x = 0.1, 0.2, 0.3$

29. Given, $dy/_{dx} = x^3 + y\,,\; y(0) = 2\,.$

The values of $y(0.2) = 2.073, y(0.4) = 2.452$, and $y(0.6) = 3.023$ are gotten by the R.K. method of the order. Find $y(0.8)$ by Milne's predictor-corrector method taking $h = 0.2$.

30. Use second-order Adams-Bashforth with $h = \frac{1}{4}$ to compute approximate values of $y(1)$ for each of the following initial value problems. Use Euler's method to generate the needed starting value.

a) $y' + 2y = 1, \;\; y(0) = 2$;

b) $y' = y(l + e^{2t}), \; y(0) = 1$;

c) $ty' = y(sint), \; y(0) = 2$;

d) $y' = y(1 - y), y(0) = \frac{1}{2}$

CHAPTER 6

Numerical Solutions of Partial Differential Equations

6.1 Partial Differential Equations: Introduction

Partial differential equations are commonly used in various fields of applied mathematics, such as hydrodynamics, elasticity, quantum physics, and electromagnetic theory. The purpose of using it is to differentiate between mathematical functions that have several variables. A solution to a partial differential equation (PDE) is a function $u(x, y, \ldots)$ that satisfies the equation throughout a certain region of the $x, y, \ldots$ variables. It is a mathematical equation that establishes a relationship between the independent variables, the dependent variable u, and its partial derivatives. It can be expressed as

$$F\left(x, y, u(x,y), u_x(x,y), u_y(x,y)\right) = F(x, y, u, u_x, u_y) = 0 \qquad (6.1)$$

This is a first-order partial differential equation in two independent variables, which is the most general form. The order of an equation is determined by the highest derivative that appears in it. The most general second-order partial differential equation in two independent variables is

$$F(x, y, u, u_x, u_y, u_{xx}, u_{yy}) = 0 \qquad (6.2)$$

When solving an ordinary differential equation (ODE), it is sometimes necessary to reverse the roles of the independent and dependent variables. For example, in the case of the separable ODE $\frac{du}{dx} = u^3$. In the context of PDEs, it is always important to maintain a clear distinction between the independent variables and the dependent variable (also known as the unknown). In partial differential equations (PDEs), we use subscripts to represent partial derivatives, such as;

$$u_x = \frac{\partial u}{\partial x}$$
$$u_{xx} = \frac{\partial^2 u}{\partial x^2}$$

$$u_{xy} = \frac{\partial^2 u}{\partial y \partial x} = \frac{\partial}{\partial y}\left(\frac{\partial u}{\partial x}\right)$$

Here are some examples of partial differential equations (PDEs) that are commonly found in physical theory:

1. $u_x + u_y = 0$ (transport)
2. $u_x + yu_y = 0$ (transport)
3. $u_x + uu_y = 0$ (shock wave)
4. $u_{xx} + u_{yy} = 0$ (Laplace's equation)
5. $u_{tt} - u_{xx} + u^3 = 0$ (wave with interaction)
6. $u_t + uu_x + u_{xxx} = 0$ (dispersive wave)
7. $u_{tt} + u_{xxxx} = 0$ (vibrating bar)
8. $u_{tt} - iu_{xx} = 0;$ $(i = \sqrt{-1})$ (quantum mechanics)

Each of these examples involves two independent variables, which can be represented as either x and y or x and t. Examples 1 to 3 have an order of one. Examples 4, 5, and 8 have an order of two. Example 6 has an order of three, and example 7 has an order of four. Examples 3, 5, and 6 stand out from the rest because they do not follow a linear pattern. Now, we will proceed to explain this concept.

Linearity refers to the concept that is being described as follows. Equation should be expressed as $\mathcal{L}u = 0$, where $\mathcal{L}$ is an operator. If v is any function, then $\mathcal{L}v$ is a new function. For instance, the operator $\mathcal{L} = \frac{\partial}{\partial x}$ takes v and computes its partial derivative with respect to x, resulting in v_x.

In Example 6.1, the operator $\mathcal{L}$ is defined as $\mathcal{L} = \frac{\partial}{\partial x} + y\frac{\partial}{\partial y}$. The equation can be rewritten as:

$$\mathcal{L}u = u_x + yu_y$$

The definition of linearity that we want to find is

$$\mathcal{L}(u+v) = \mathcal{L}u + \mathcal{L}v; \qquad \mathcal{L}(cu) = c\mathcal{L}u \tag{6.3}$$

for any functions u, v and any constant C. Whenever (6.3) holds (for all choices of u, v, and C), l is called linear operator. The equation

$$\mathcal{L}u = 0 \tag{6.4}$$

is called linear when $\mathcal{L}$ is a linear operator. r. Equation (6.4) is called a homogeneous linear equation. The equation

$$\mathcal{L}u = g \tag{6.5}$$

where $g \neq 0$ is a given function of the independent variables, is called an inhomogeneous linear equation. For example, the equation

$$(\cos xy^2)u_x - y^2u_y = \tan(x^2 + y^2) \tag{6.6}$$

is an inhomogeneous linear equation.

We will primarily focus on studying linear systems with constant coefficients. It is important to remember that when dealing with ordinary differential equations (ODEs), you can obtain solutions through linear combinations. The coefficients represent the arbitrary constants. In the case of an ordinary differential equation (ODE) of order m, there are m arbitrary constants that can be determined.

Let's have a look at some PDEs.

Example 6.1 Determine all values of $u(x, y)$ that satisfy $u_{xx} = 0$. To obtain the value of u_x, we can integrate once, resulting in a constant value. However, that statement is not entirely accurate because there is another variable, y, to consider. The equation we have is $u_x(x, y) = f(y)$, where $f(y)$ can be any arbitrary function.

The process is to be repeated to obtain the function $u(x, y) = f(y)x + g(y)$. This is the formula for the solution. It is noted that the solution includes two arbitrary functions. We also observe this in the following two examples.

Example 6.2 Solve the partial differential equation (PDE) given by the equation $u_{xx} + u = 0$. Once again, it can be observed that the given equation is an ordinary differential equation (ODE) with an additional variable, denoted as y. The ODE can be solved, resulting in the solution: The solution can be expressed as

$$u = f(y)\cos x + g(y)\sin x$$

where $f(y)$ and $g(y)$ are arbitrary functions of y. One can readily verify the accuracy of this formula by performing two differentiations to confirm that $u_{xx} = -u$.

Example 6.3 Solve the partial differential equation (PDE) $u_{xy} = 0$. This task is not too difficult either. First, let's perform integration with respect to x, while considering y as a constant. Thus, we obtain

$$u_y(x, y) = f(y)$$

Next, let us proceed with the integration of y with respect to x, while considering x as a constant. We obtain the solution

$$u(x, y) = F(y) + G(x),$$

Where $F' = f$.

6.2 Classification of Partial Differential Equations (PDE)

Each type of partial differential equation (PDE) possesses certain characteristics which help in determining whether it is appropriate of a finite element approach for solving the problem described by the PDE. The solution is dependent on the equation and involves several variables that include partial derivatives with respect to those variables.

PDEs can be classified into three categories:

a) Hyperbolic PDE
b) Parabolic PDE and
c) Elliptic PDEs.

From a physical perspective, these partial differential equations (PDEs) correspondingly represent wave propagation, time-dependent diffusion processes, and steady state or equilibrium processes. Hyperbolic equations are utilized to represent the transportation of various physical quantities, such as fluids or waves. Parabolic problems are used to describe evolutionary phenomena that ultimately reach a stable state, which can be mathematically described by an elliptic equation. Elliptic equations are linked to a distinct state within a system, which theoretically corresponds to the energy's minimum.

From a mathematical standpoint, the classification of second-order partial differential equations (PDEs) is determined by the potential to transform equation into canonical or standard form at a specific point through coordinate transformation. It is important to note that, when classifying, there is no need to limit the consideration solely to linear equations. This also applies to quasilinear second-order partial differential equations. A quasilinear second-order partial differential equation (PDE) is characterized by its linearity solely in the second derivatives.

Consider this example

$$au_{xx} + bu_{yy+}cu_{yy} = 0, \quad u = u(x, y)$$

For a given point (x, y), the equation is said to be **Elliptic** if $b^2 - ac < 0$ which are used to explain the equations of elasticity without inertial terms. **Hyperbolic** PDEs describe the phenomena of wave propagation if it satisfies the certain condition $b^2 - ac > 0$. **Parabolic** partial differential equations (PDEs) must satisfy the condition $b^2 - ac = 0$. The heat conduction equation is a parabolic partial differential equation (PDE).

6.3 Solutions of Parabolic Equation

A parabolic partial differential equation is a particular class of partial differential equations (PDEs). Parabolic partial differential equations (PDEs) are used for modeling wide time-dependent phenomena such as heat conduction, particle diffusion, and the valuation of derivative investment instruments.

The equation

$$u_{xx} = u_t \tag{6.7}$$

represents the temperature distribution along a thin rod over time. The solutions to this problem, although seemingly simple, are intricate and primarily rely on a function known as the fundamental solution of the equation. This function is an exponential function expressed as

$$e^{[(-x2/4t)/t1/2]} \tag{6.8}$$

To obtain the comprehensive solution for this problem, it is necessary to have knowledge of the initial temperature distribution along the rod as well as the rate at which the temperature at the ends of the rod is changing. The conditions referred to as initial values and boundary values are collectively known as auxiliary conditions.

In both the two- and three-dimensional problems, it is necessary to have knowledge of the initial temperature distribution within the region, as well as the temperature distribution along the boundary at each moment. The basic form of the differential equation in two dimensions is $u_{xx} + u_{yy} = u_t$. In the three-dimensional case, an additional u_{zz} term is included. These equations are applicable only in cases where the medium has a uniform composition. However, in situations involving nonuniform composition or other diffusion-type problems, more complex equations may be necessary. The equations in the given region are referred to as parabolic if they can be expressed in a simpler form using an alternative coordinate system. If the discriminant $b^2 - 4ac$ of a one-dimensional equation is equal to zero, the equation can be transformed in a way that the higher-order terms become $au_{xx} + bu_{xt} + cu_{tt}$. The equation will exhibit a parabolic shape within a region if the coefficients a, b, and c are dependent on the values of x, and if the expression $b^2 - 4ac = 0$ at every point within that region.

The following one-dimensional (1D) differential equation

$$u_t - a^2 u_{xx} = f(x,t), \qquad x \in (0, L) \tag{6.9}$$

represents a one-dimensional (1D) parabolic equation, commonly known as the heat/diffusion equation.

The equation is used to represent various physical phenomena, including the distribution of heat on a rod. In this equation, $u(x,t)$ represents the temperature at a specific position x and time t. The parameter a^2 represents the thermal diffusivity of the material, and $f(x,t)$ represents a source or sink along the rod.

The coefficient α varies with position x when the material property is non-uniform along the rod. The thermal conductivity, K, in this scenario is dependent on the position, x. Consequently, the heat equation can be expressed as

$$u_t - \nabla \cdot (K(x)\, u_x)_x = f(x,t) \qquad (6.10)$$

6.3.1 The Explicit Method

Equations such as

$$\frac{\partial T}{\partial} = \propto \frac{\partial^2 T}{\partial x^2} \qquad (6.11)$$

represent parabolic partial differential equations (PDEs).

In this context, the variables are defined as follows: T represents temperature, t represents time, α is the thermal diffusivity (defined as $\propto = k/\rho C_p$), k is the thermal conductivity, ρ is the density, and C_p is the heat capacity. The given equation represents the time-dependent temperature distribution in a one-dimensional body, which is modeled as a slender rod in the provided sketch. In this context, the length of the rod in the x-direction is significantly larger than its typical cross-sectional dimensions. The heat conduction equation can be verified to be parabolic.

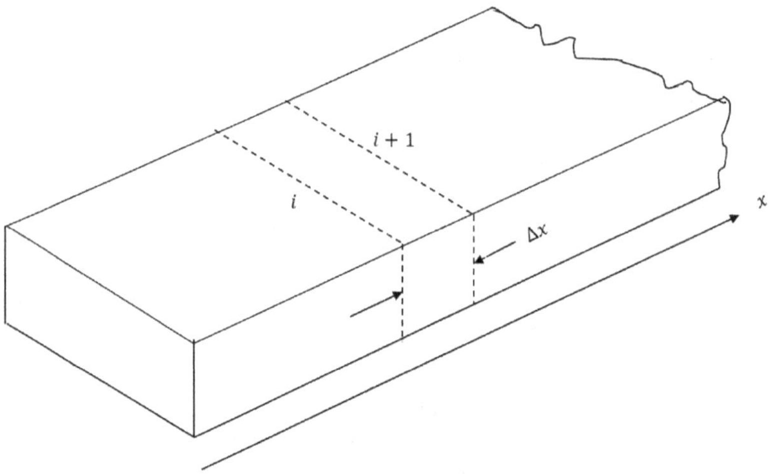

Figure 6.1 Time-Dependent Temperature Distribution in a One-Dimensional Body

The equation is a one-dimensional space equation, as it does not depend on y or z. However, it is two-dimensional in that it includes the dependence on time. However, it is important to note that the computational domain in the examples discussed in the previous lecture is not two-dimensional. The absence of a time constraint is the reason for this. The bar can be divided into nodes, spaced x units apart, and labeled as i in these Notes. It is beneficial to implement a comparable numbering system for the "nodes" in time, utilizing the variable j. We aim to determine $T(x_i, t_{j+1})$ based on our knowledge of $T(x_i, t_j)$ for all i and j. The time derivative can be approximated using forward differences as

$$\frac{\partial T}{\partial}(x_i, t_j) \approx \frac{T(x_i, t_{j+i}) - T(x_i, t_j)}{\Delta t} + O(\Delta t)$$

And change notation to read

$$\frac{\partial T}{\partial} \approx \frac{T_i^{j+1} - T_i^j}{\Delta t} + O(\Delta t) \tag{6.12}$$

The use of superscripts to indicate time nodes is prevalent. It enables the inclusion of additional subscripts for spatial problems with multiple dimensions. By central differences we get,

$$\frac{\partial^2 T}{\partial x^2} \approx \frac{T_{i+1}^j - 2T_i^j + T_{i-1}^j}{(\Delta x)^2} + O(\Delta x)^2 \tag{6.13}$$

It is important to note that Eq (6.12) represents the forward difference formula, which is considered to be less accurate compared to Eq (6.13), which represents the central difference formula. These equations when combined provide

$$T_i^{j+1} = \frac{\propto \Delta t}{(\Delta x)^2}\left(T_{i+1}^j + T_{i-1}^j\right) + \left(1 - \frac{2\propto \Delta t}{(\Delta x)^2}\right) T_i^j \tag{6.14}$$

so that we have an explicit equation for conditions at time T_{j+1} from the previously calculated conditions at t_j. By defining

$$r = \frac{\propto \Delta t}{(\Delta x)^2}$$

the equation becomes

$$T_i^{j+1} = r\left(T_{i+1}^j + T_{i-1}^j\right) + (1 - 2r)T_i^j \tag{6.15}$$

6.4 Implicit Methods

These methods are more stable compared to the explicit methods discussed earlier. One important example that demonstrates the formulation of implicit methods is the Crank-Nicholson (CN) formulation. This formulation is based on the observation that Equation (6.12) represents a forward difference at time t_j, but a central difference at time $t_{j+\frac{1}{2}}$, which is halfway between t_j and t_{j+1}.

If we consider the central difference formula for $\frac{\partial^2 T}{\partial x^2}$ at t_j, Equation (6.13) represents the expression for T divided by x squared at time t and position j. Additionally, a similar equation is required for $\frac{\partial^2 T}{\partial x^2}$ at t_{j+1}. When dividing this sum by 2, the result is

$$\frac{T_i^{j+1} - T_i^j}{\Delta t} = \propto [\frac{T_{i+1}^j - 2T_i^j + T_{i-1}^j}{2(\Delta x)^2} + \frac{T_{i+1}^{j+1} - 2T_i^{j+1} + T_{i-1}^{j+1}}{2(\Delta x)^2}$$

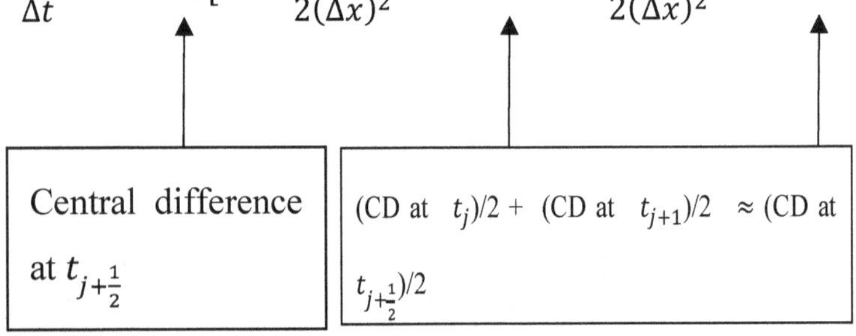

This procedure ensures that the time and spatial derivatives have the same order of accuracy. The equation corresponding to equation (6.14) is

$$T_i^{j+1} = T_i^j + \frac{r}{2}[T_{i+1}^j - 2T_i^j + T_{i-1}^j + T_{i+1}^{j+1} - 2T_i^{j+1} + T_{i-1}^{j+1}]$$

$$\text{i.e., } (1+r)T_i^{j+1} = (1-r)T_i^j + \frac{r}{2}[T_{i+1}^j + T_{i-1}^j + T_{i+1}^{j+1} + T_{i-1}^{j+1}] \quad (6.16)$$

The determination of T_i^{j+1} must be done implicitly due to its equation's dependence on other temperatures at t_{j+1}. In other words, Equation (6.16) can be reformulated as follows:

$$-\frac{r}{2}T_{i-1}^{j+1} + (1+r)T_i^{j+1} - \frac{r}{2}T_{i+1}^{j+1} = \frac{r}{2}T_{i-1}^{j} + (1-r)\frac{r}{2}T_i^{j} + \frac{r}{2}T_{i-1}^{j} = b_i^{j+1} \quad (6.17)$$

b_i^{j+1} is known after the jth iteration so we have a set of tridiagonal linear equations for the temperatures at t_{j+1}.

Example 6.4 Consider the equation $\frac{\partial T}{\partial} = \propto \frac{\partial^2 T}{\partial x^2}$ for a thin rod of length 1 unit. If T is understood as the temperature above ambient, and the initial temperature distribution is

$$T = T_0 \sin \pi x$$

Using $r = \propto \frac{\Delta t}{\Delta x^2}$ as before, the exact solution is given by

$$\frac{T}{T_0} = \sin \pi x e^{-j\pi^2 r/16} \quad (6.18)$$

where$j - 1,2,3,. . .$ are the numbers of the time nodes. By dividing the rod into lengths $\Delta x = 0.25$, solve Equations (6.15) and (6.17) for $r = \frac{1}{2}$ and 1 and compare the results to the exact solution.

Solution Explicit Solutions (using Eq. 6.15)

$$T_i^{j+1} = (T_{i+1}^{j} + T_{i-1}^{j})/2 \quad (6.19)$$

which is perhaps the simplest form of Eq (6.15) - apart from $r = 0$. The equation is almost trivial. Note the following initial and symmetry conditions:

Chart: 6a

i	x	$T(x,0)/T_0$
0	0	0
1	0.25	0.70711
2	0.50	1
3	0.75	0.70711

And $T_0{}^j = 0$ for all j and $T_3{}^j = T_1{}^j$

Using Eq (6.19) we can construct a table of $T_i^{\,j}/T_0$ using $T_1^{j+1} = T_2^j/2$ and $T_2^{j+1} = T_1^j$.

Table 6.1 $T_i^{\,j}/T_0$ and errors for explicit solution and $r = 0$

	$i = 0$	$\Delta(0,j)$	$i = 1$	$\Delta(1,j)$	$i = 2$	$\Delta(2,j)$
$j = 0$	0	0	0.7071	0	1000	0
$j = 1$	0	0	0.5	3.7	0.7071	3.7
$j = 2$	0	0	0.3536	7.3	0.5000	7.3
$j = 3$	0	0	0.2500	10.8	0.3536	10.814.2
$j = 4$	0	0	0.1768	14.2	0.2500	14.2
$j = 5$	0	0	0.1250	17.3	0.1768	17.3
$j = 6$	0	0	0.0884	20.5	0.1250	20.5

The formula for the relative error Δ is:

$$\Delta = \left| \frac{T_i^j - \dfrac{T}{T_0}}{T_i^j} \right| \%$$

Now,

$r = 1$

$$T_i^{j+1} = T_{i+1}^j + T_{i-1}^j - T_i^j$$
$$\text{So, } T_1^{j+1} = T_2^j + T_1^j$$
$$T_2^{j+1} = 2T_1^j - T_2^j$$

Table 6.2 $T_i^{\,j}/T_0$ and errors for explicit solution and $r = 1$

	$i = 0$	$\Delta(0,j)$	$i = 1$	$\Delta(1,j)$	$i = 2$	$\Delta(2,j)$
$j = 0$	0	0	0.7071	0	1000	0
$j = 1$	0	0	0.2929	23.2	0.4142	23.2
$j = 2$	0	0	0.1213	41.1	0.1716	41.1
$j = 3$	0	0	0.0503	55.0	0.0711	55.0

Note: that the value of j in this calculation is equivalent (in real time) to $2j$ in the previous. The relative error Δ is defined as previously.
Explicit Solutions (using Eq. 6.17)

$r = 1/2$

$$\frac{3}{2}T_i^{j+1} - \frac{1}{4}T_{i-1}^{j+1} - \frac{1}{4}T_{i+1}^{j+1} = \frac{1}{2}T_i^j + \frac{1}{4}T_{i+1}^j + \frac{1}{4}T_{i-1}^j$$

For $i = 1$:

$$\frac{3}{2}T_1^{j+1} - \frac{1}{4}T_2^{j+1} = \frac{1}{2}T_1^j + \frac{1}{4}T_2^j$$

For $i = 2$:

$$\frac{3}{2}T_2^{j+1} - \frac{1}{2}T_1^{j+1} = \frac{1}{2}T_2^j + \frac{1}{2}T_1^j$$

Thus,

$T_1^{j+1} = (7T_1^j + 4T_2^j)/17$ and $T_2^{j+1} = (7T_2^j + 8T_1^j)/17$

which are slightly more complicated than the equations for the explicit solution.

We can construct a similar table for $T_i{}^j/T_0$.

Table 6.3 $T_i{}^j/T_0$ and errors for explicit solution and $r = 1/2$

	$i = 0$	$\Delta(0,j)$	$i = 1$	$\Delta(1,j)$	$i = 2$	$\Delta(2,j)$
$j = 0$	0	0	0.7071	0	1.000	0
$j = 1$	0	0	0.5242	1.0	0.7420	1.0
$j = 2$	0	0	0.3904	2.3	0.5522	2.3
$j = 3$	0	0	0.2907	3.7	0.4111	3.7
$j = 4$	0	0	0.2164	5.1	0.3061	5.1
$j = 5$	0	0	0.1611	5.1	0.2279	5.1
$j = 6$	0	0	0.1200	8.0	0.2697	8.0

$r = 1$

Eq (6.17) gives

$$2T_i^{j+1} - \frac{1}{2}T_{i-1}^{j+1} - \frac{1}{2}T_{i+1}^{j+1} = \frac{1}{2}T_{i+1}^j + \frac{1}{2}T_{i-1}^j$$

So that $T_1^{j+1} = \frac{2}{7}T_2^j + \frac{1}{7}T_1^j$

$$T_2^{j+1} = \frac{1}{7}T_2^j + \frac{4}{7}T_1^j$$

Table 6.4 T_i^j/T_0 and errors for explicit solution and $r = 1$

	$i = 0$	$\Delta(0,j)$	$i = 1$	$\Delta(1,j)$	$i = 2$	$\Delta(2,j)$
$j = 0$	0	0	0.7071	0	1000	0
$j = 1$	0	0	0.3867	1.34	0.5469	1.34
$j = 2$	0	0	0.2115	2.7	0.2991	2.7
$j = 3$	0	0	0.1157	4.1	0.1636	4.1

The value of j for $r = 1$ corresponds to the real time of $2j$ to $r = 1/2$. In this particular example, the explicit solution remains stable when $r = 1$. However, there is a noticeable decrease in accuracy as the value of r increases. The implementation of the implicit Crank-Nicholson solution is more challenging due to the increased complexity of deriving and solving the equations for T_i^{j+1}. However, it offers a significant improvement in accuracy. The implicit solution for $r = 1$ is more accurate than the implicit solution for $r = 1/2$. The accuracy at each step is approximately equal, but the case where $r = 1$ corresponds to a doubling of the step in real time. I am uncertain about the underlying cause for this phenomenon. As the number of spatial nodes increased, the equations became too numerous to solve directly. Eq (6.16) can be written as

$$\begin{bmatrix} 1+r & -r/2 & 0 & 0 & . \\ -r/2 & 1+r & -r/2 & 0 & . \\ . & . & . & . & . \\ . & . & . & . & . \\ 0 & 0 & 0 & -r/2 & 1+r \end{bmatrix} \begin{bmatrix} T_1^{j+1} \\ T_2^{j+1} \\ T_3^{j+1} \\ . \\ T_n^{j+1} \end{bmatrix} = \begin{bmatrix} b_1^{j+1} \\ b_2^{j+1} \\ b_3^{j+1} \\ . \\ b_n^{j+1} \end{bmatrix} \tag{6.20}$$

The terms in question, denoted as b^{j+1}, are calculated based on the temperatures at time t^j. The coefficient matrix is both tridiagonal and diagonally dominant for any positive value of r. Therefore, the LU factorization and substitution method used in Example 6.3.2.1 can be easily applied.

6.5 Stability

The explicit method is often found to be unstable. Typically, a value of r less than $1/2$ is required for stability and convergence. Stability refers to the property where errors occurring at any given point do not result in exponentially growing errors for subsequent points. Convergence refers to the capability of a numerical solution to approach the analytic solution, assuming it exists, as the values of x and t decrease. A comprehensive stability analysis typically involves the determination of eigenvalues. The analysis demonstrates that the CN and other implicit methods exhibit stability across all values of r.

6.6 Elliptic Equations

Elliptic equations are a type of partial differential equations that describe phenomena that remain constant over time, such as the flow of heat or fluid in a medium without any accumulations. For an elliptic partial differential equation, the discriminant $\Delta = b^2 - ac < 0$. The Laplace equation, $u_{xx} + u_{yy} = 0$, is the most basic equation that describes this condition in two dimensions. Elliptic equations are commonly employed in engineering to describe steady-state boundary-value problems.

The Laplace equation $\Delta^2 u = \frac{\partial^2 u}{\partial x^2} + \frac{\partial^2 u}{\partial y^2} = 0$ and Poisson's equation $\frac{\partial^2 u}{\partial x^2} + \frac{\partial^2 u}{\partial y^2} = f(x,y)$ are both examples of elliptic partial differential equations. The Laplace equation is important in situations involving steady-state flow and potential problems.

Poisson's equation is used in the fields of fluid mechanics, electricity and magnetism, and torsion problems.

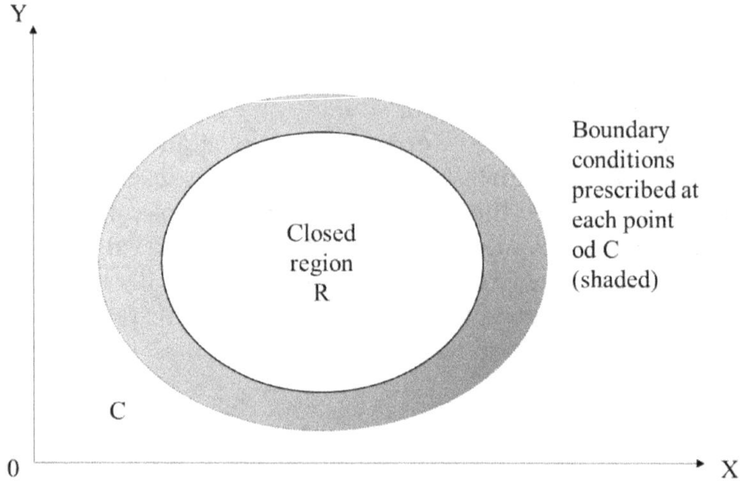

Figure 6.2 Boundary Condition

The equations have a solution in the form of a function $u(x, y)$ that satisfies the specified boundary conditions on the closed curve C within the region R (Fig 6-2).

Typically, issues related to steady viscous flow and equilibrium stresses in elastic structures involve elliptic equations.

6.7 Solution of Laplace's Equation

The Laplace equation can be used to model a wide range of problems that involve the possibility of an unknown variable. Our primary setting for deriving and solving this elliptic PDE will be a heated plate due of its simplicity and general relevance to most disciplines of engineering.

$$\frac{\partial^2 u}{\partial x^2} + \frac{\partial^2 u}{\partial y^2} = 0 \tag{6.21}$$

Let R be a rectangular region with known boundary values of $u(x, y)$. The region should be divided into a network of square meshes with a side length of h, as depicted in Figure 6-3, assuming that a precise subdivision of R is feasible. Replacing the derivatives in (6.21) by their difference approximations, we have

$$\frac{1}{h^2}\left[u_{i-1,j} - 2u_{i,j} + u_{i+1,j}\right] + \frac{1}{h^2}\left[u_{i,j-1} - 2u_{i,j} + u_{i,j+1}\right] = 0$$

Or $$u_{i,j} = \frac{1}{4}\left[u_{i-1,j} + u_{i+1,j} + u_{i,j+1} + u_{i,j-1}\right] \qquad (6.22)$$

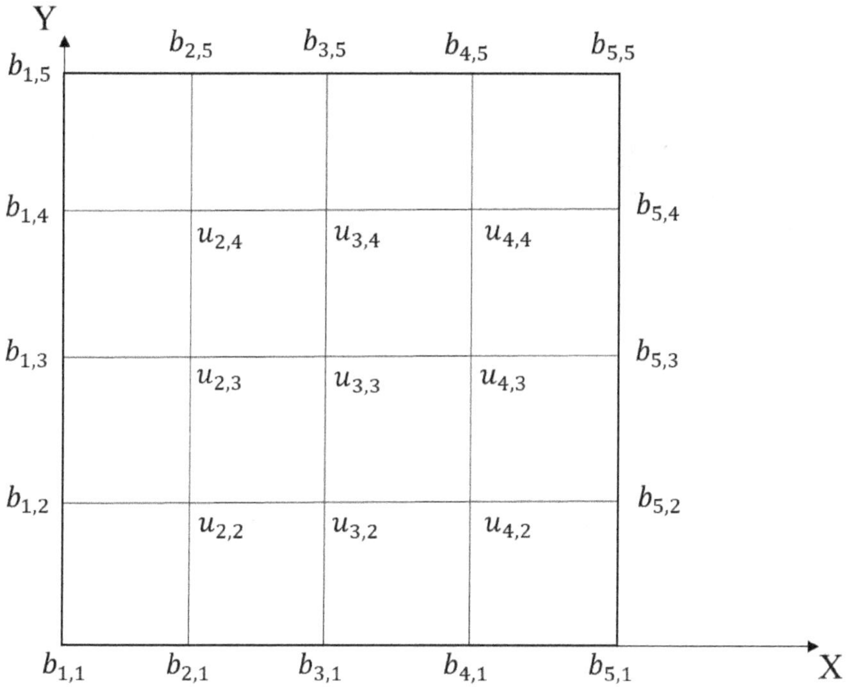

Figure 6.3 Interior mesh points and boundary points

This indicates that the value of u at any internal mesh point is determined by taking the average of its values at four adjacent points: to the left, right, above, and below. The **standard 5-point formula**, shown in Figure 6-4, is referred to as (6.22).

Sometimes a formula similar to (6.22) is used which is given by

$$u_{i,j} = \frac{1}{4}\left(u_{i-1,j} + u_{i+1,j} + u_{i+1,j+1} + u_{i-1,j-1}\right) \qquad (6.23)$$

This shows that the value of $u_{i,j}$ is determined by taking the average of its values at the four neighboring diagonal mesh points. The formula represented in Figure 6-4 is known as the diagonal five-point formula, denoted as (6.23). While (6.23) is less precise than (6.22), it still serves as a reasonably effective method for obtaining the initial values at the mesh points.

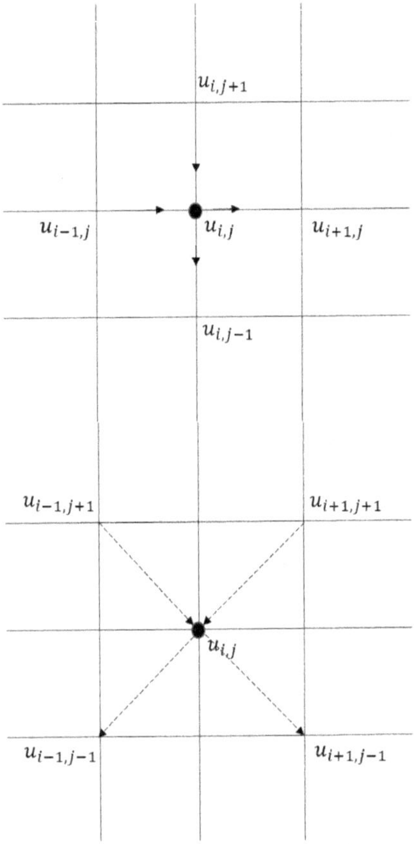

Figure 6.4 Approximations to Laplace's Equation

In order to determine the initial values of u at the inner mesh points, we first compute $u_{3,3}$, $u_{2,4}$, $u_{4,4}$, $u_{4,2}$ and $u_{2,2}$ in that order, using the diagonal five-point formula (6.23). Thus, we obtain,

$$u_{3,3} = \frac{1}{4}\left(b_{1,5} + b_{5,1} + b_{5,5} + b_{1,1}\right)$$

$$u_{2,4} = \frac{1}{4}\left(b_{1,5} + b_{3.3} + b_{3,5} + b_{1,3}\right)$$

$$u_{4,4} = \frac{1}{4}\left(b_{3,5} + b_{5,3} + b_{3,5} + b_{3,3}\right)$$

$$u_{3,3} = \frac{1}{4}\left(b_{1,5} + b_{5,1} + b_{5,5} + b_{1,1}\right)$$

$$u_{2,2} = \frac{1}{4}(b_{1,3} + b_{3,1} + b_{3,3} + b_{1,1})$$

The values at the remaining interior points, i.e., $u_{2,3}$, $u_{3,4}$, $u_{3,4}$, $u_{4,3}$ and $u_{3,2}$ are computed by the standard five-point formula (6.22). Thus, we obtain

$$u_{2,3} = \frac{1}{4}(b_{1,3} + b_{3,3} + b_{2,4} + b_{2,2})$$
$$u_{3,4} = \frac{1}{4}(b_{2,4} + b_{4,4} + b_{3,5} + b_{3,3})$$
$$u_{4,3} = \frac{1}{4}(b_{3,3} + b_{5,3} + b_{4,4} + b_{4,2})$$
$$u_{3,2} = \frac{1}{4}(b_{2,2} + b_{4,2} + b_{3,3} + b_{3,1})$$

After obtaining all nine values of $u_{i,j}$ once, their accuracy is enhanced using one of the iterative approaches listed below. The process is repeated until the difference between two consecutive iterations becomes minimal in each case.

1. **Jacobi's method.** Denoting the nth iterative value of $u_{i,j}$, by ${u^n}_{i,j}$, the iterative formula to solve (6.22) is

$$u_{i,j}^{(n+1)} = \frac{1}{4}\left[{u^n}_{i-1,j} + {u^n}_{i+1,j} + {u^n}_{i,j+1} + {u^n}_{i,j-1}\right] \tag{6.24}$$

It is known as the point Jacobi's formula as it improves the values of $u_{i,j}$ at the interior mesh points.

2. **Gauss-Seidal method**. The Guass-Seidel method is a modified version of the Jacobi method. The value of the variables in the Jacobi method is not changed until the following iteration, but the value of the variables in the Gauss-Seidel technique is changed as soon as a new value is determined. The iteration formula for this method is

$$u_{i,j}^{(n+1)} = \frac{1}{4}\left[{u^{(n+1)}}_{i-1,j} + {u^{(n)}}_{i+1,j} + {u^{(n+1)}}_{i,j+1} + {u^{(n)}}_{i,j-1}\right]$$

It uses the most recent iteration value and scans the mesh points symmetrically from left to right along successive rows.

Example 6.5 For the following square mesh with boundary values illustrated in Fig. 6-5, solve the elliptic equation $u_{xx} + u_{yy} = 0$.

Solution. Let $u_1, u_2, \ . \ . \ . \ u_9$ be the values of u at the interior mesh-points. Since the boundary values of u are symmetrical about AB,

$$\therefore \ u_7 = u_1,$$
$$u_8 = u_2,$$
$$u_9 = u_3$$

And the values of u being symmetrical about CD.

$u_3 = u_1, u_6 = u_4, u_9 = u_7$

Thus, it is sufficient to find the values u_1, u_2, u_4 and u_5.

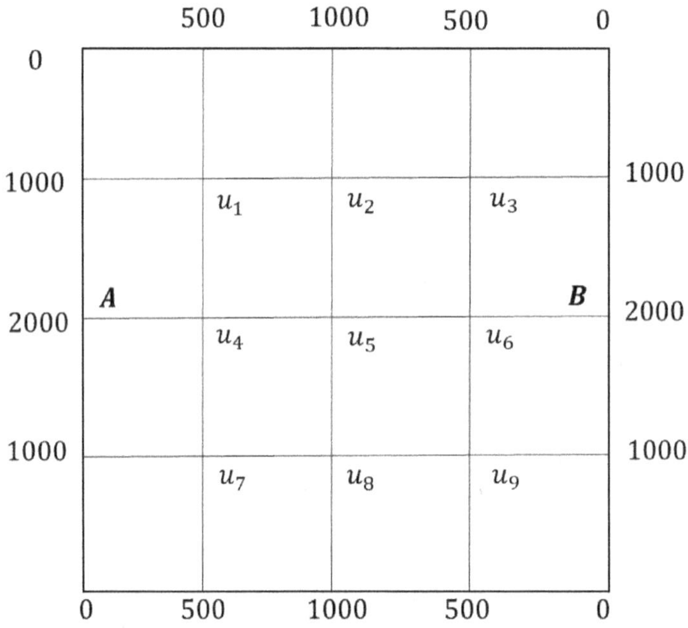

Figure 6.5 Square mesh for example 6.5.1

Now we find their initial values in the following order:

$$u_5 = \frac{1}{4}(2000 + 2000 + 1000 + 1000) = 1500 \qquad \text{(Std. formula)}$$

$$u_1 = \frac{1}{4}(0 + 1500 + 1000 + 2000) = 1125 \qquad \text{(Diag. formula)}$$

$$u_2 = \frac{1}{4}(1125 + 1125 + 1000 + 1500) \approx 1188 \qquad \text{(Std. formula)}$$

$$u_4 = \frac{1}{4}(2000 + 1500 + 1125 + 1125) \approx 1438 \qquad \text{(Std. formula)}$$

Now the iteration process is carried out using the standard formulae:

$$\begin{aligned}
u_1^{(n+1)} &= \frac{1}{4}\left[1000 + u_2^{(n)} + 500 + u_4^{(n)}\right] \\
u_2^{(n+1)} &= \frac{1}{4}\left[u_1^{(n+1)} + u_1^{(n)} + 1000 + u_5^{(n)}\right] \\
u_4^{(n+1)} &= \frac{1}{4}\left[2000 + u_5^{(n)} + u_1^{(n+1)} + u_1^{(n)}\right] \\
u_5^{(n+1)} &= \frac{1}{4}\left[u_4^{(n+1)} + u_4^{(n)} + u_2^{(n+1)} + u_2^{(n)}\right]
\end{aligned}$$

First iteration: (put $n = 0$ in the above results)

$$\begin{aligned}
u_1^{(1)} &= \frac{1}{4}(1000 + 1188 + 500 + 1438) = 1032 \\
u_2^{(1)} &= \frac{1}{4}(1032 + 1125 + 1000 + 1500) = 1164 \\
u_4^{(1)} &= \frac{1}{4}(2000 + 1500 + 1032 + 1125) = 1414 \\
u_5^{(1)} &= \frac{1}{4}(1414 + 1438 + 1164 + 1188) = 1301
\end{aligned}$$

Second iteration: (put $n = 1$)

$$\begin{aligned}
u_1^{(2)} &= \frac{1}{4}(1000 + 1164 + 500 + 1414) = 1020 \\
u_2^{(2)} &= \frac{1}{4}(1020 + 1032 + 1000 + 1301) = 1088 \\
u_4^{(2)} &= \frac{1}{4}(2000 + 1301 + 1020 + 1032) = 1338 \\
u_5^{(2)} &= \frac{1}{4}(1338 + 1414 + 1088 + 1164) = 1251
\end{aligned}$$

Third iteration:

$$u_1^{(3)} = \frac{1}{4}(1000 + 1088 + \ 500 + 1338) = 982$$
$$u_2^{(3)} = \frac{1}{4}(982 + 1020 + \ 1000 + 1251) = 1063$$
$$u_4^{(3)} = \frac{1}{4}(2000 + 1251 + \ 982 + 1020) = 1313$$
$$u_5^{(3)} = \frac{1}{4}(1313 + 1338 + \ 1063 + 1088) = 1201$$

Fourth iteration:

$$u_1^{(4)} = \frac{1}{4}(1000 + 1063 + \ 500 + 1313) = 969$$
$$u_2^{(4)} = \frac{1}{4}(969 + 982 + \ 1000 + 1201) = 1038$$
$$u_4^{(4)} = \frac{1}{4}(2000 + 1201 + \ 969 + 982 = 1288$$
$$u_5^{(4)} = \frac{1}{4}(1288 + 1313 + \ 1038 + 1063) = 1176$$

A slight difference between the values obtained in the third and fourth iterations is observed. Additional iterations are expected to bring about further improvement in the accuracy of the computed values.

Example 6.6 Given the values of $u(x, y)$ on the boundary of the square in the Fig. 6-6, evaluate the function $u(x, y)$ satisfying the Laplace equation $\Delta^2 u = 0$ at the pivotal points of this figure by

a) Jacobi's method
b) Gauss-Seidal method

Solution. To obtain the initial values of u_1, u_2, u_3, u_4, we assume that $u_4 = 0$. Then,

$$u_1 = \frac{1}{4}(1000 + 0 + 1000 + 2000) = 1000 \qquad \text{(Diag. formula)}$$
$$u_2 = \frac{1}{4}(1000 + 500 + 1000 + 0) \approx 625 \qquad \text{(Std. formula)}$$

$$u_3 = \frac{1}{4}(2000 + 0 + 1000 + 500) \approx 875 \qquad \text{(Std. formula)}$$
$$u_4 = \frac{1}{4}(875 + 0 + 625 + 0) \approx 375 \qquad \text{(Std. formula)}$$

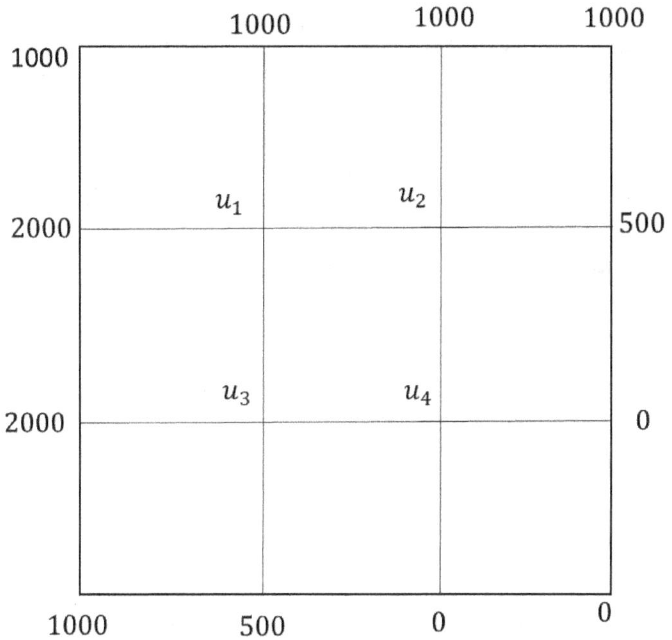

Figure 6.6 Square mesh for example 6.6

(a) Using Jacobi's formulae, the successive iterations:

$$u_1^{(n+1)} = \frac{1}{4}\left[2000 + u_2^{(n)} + 1000 + u_3^{(n)}\right]$$
$$u_2^{(n+1)} = \frac{1}{4}\left[u_1^{(n)} + 500 + 1000 + u_4^{(n)}\right]$$
$$u_3^{(n+1)} = \frac{1}{4}\left[2000 + u_4^{(n)} + u_1^{(n)} + 500\right]$$
$$u_4^{(n+1)} = \frac{1}{4}\left[u_3^{(n)} + 0 + u_2^{(n)} + 0\right]$$

First iteration: (put $n = 0$ in the above results)

$$u_1^{(1)} = \frac{1}{4}(2000 + 625 + 1000 + 875) = 1125$$
$$u_2^{(1)} = \frac{1}{4}(1000 + 500 + 1000 + 375) = 719$$

$$u_3^{(1)} = \frac{1}{4}(2000 + 375 + \ 1000 + 500) = 969$$

$$u_4^{(1)} = \frac{1}{4}(875 + 0 + 625 + 0) = 375$$

Second iteration: (put $n = 1$)

$$u_1^{(2)} = \frac{1}{4}(2000 + 719 + 1000 + 969) = 1172$$

$$u_2^{(2)} = \frac{1}{4}(1125 + 500 + 1000 + 375) = 750$$

$$u_3^{(2)} = \frac{1}{4}(2000 + 375 + 1125 + 500) = 1000$$

$$u_4^{(2)} = \frac{1}{4}(969 + 0 + 719 + 0) = 422$$

Similarly,

$u_1^{(3)} \approx 1188,\ u_2^{(3)} \approx 774,\ u_3^{(3)} \approx 1024,\ u_4^{(3)} \approx 438$
$u_1^{(4)} \approx 1200,\ u_2^{(4)} \approx 782,\ u_3^{(4)} \approx 1032,\ u_4^{(4)} \approx 450$
$u_1^{(5)} \approx 1204,\ u_2^{(5)} \approx 788,\ u_3^{(5)} \approx 1038,\ u_4^{(5)} \approx 454$
$u_1^{(6)} \approx 1206,\ u_2^{(6)} \approx 790,\ u_3^{(6)} \approx 1040,\ u_4^{(6)} \approx 456$
$u_1^{(3)} \approx 1208,\ u_2^{(3)} \approx 791,\ u_3^{(3)} \approx 1041,\ u_4^{(3)} \approx 458$
$u_1^{(3)} \approx 1208,\ u_2^{(3)} \approx 791.5,\ u_3^{(3)} \approx 1041.5,\ u_4^{(3)} \approx 458$
There is no significant difference between the seventh and eighth iteration values.

Hence $u_1 = 1208,\ u_2 = 791.5,\ u_3 = 1041.5,\ u_4 = 458$

(b) Using Gauss-Seidal formulae, the successive iterations:

$$u_1^{(n+1)} = \frac{1}{4}\left[2000 + u_2^{(n)} + 1000 + u_3^{(n)}\right]$$

$$u_2^{(n+1)} = \frac{1}{4}\left[u_1^{(n+1)} + 500 + 1000 + u_4^{(n)}\right]$$

$$u_3^{(n+1)} = \frac{1}{4}\left[2000 + u_4^{(n)} + u_1^{(n+1)} + 500\right]$$

$$u_4^{(n+1)} = \frac{1}{4}\left[u_3^{(n+1)} + 0 + u_2^{(n+1)} + 0\right]$$

First iteration: (put $n = 0$ in the above results)

$$u_1^{(1)} = \frac{1}{4}(2000 + 625 + 1000 + 875) = 1125$$
$$u_2^{(1)} = \frac{1}{4}(1125 + 500 + 1000 + 375) = 750$$
$$u_3^{(1)} = \frac{1}{4}(2000 + 375 + 1125 + 500) = 1000$$
$$u_4^{(1)} = \frac{1}{4}(1000 + 0 + 750 + 0) = 438$$

Second iteration: (put $n = 1$)

$$u_1^{(2)} = \frac{1}{4}(2000 + 750 + 1000 + 1000) = 1188$$
$$u_2^{(2)} = \frac{1}{4}(1188 + 500 + 1000 + 438) = 782$$
$$u_3^{(2)} = \frac{1}{4}(2000 + 438 + 1188 + 500) = 1032$$
$$u_4^{(2)} = \frac{1}{4}(1032 + 0 + 782 + 0) = 454$$

Similarly,
$u_1^{(3)} \approx 1204,\ u_2^{(3)} \approx 789,\ u_3^{(3)} \approx 1040,\ u_4^{(3)} \approx 458$
$u_1^{(4)} \approx 1207,\ u_2^{(4)} \approx 791,\ u_3^{(4)} \approx 1041,\ u_4^{(4)} \approx 458$
$u_1^{(5)} \approx 1208,\ u_2^{(5)} \approx 791.5,\ u_3^{(5)} \approx 1041.5,\ u_4^{(5)} \approx 458.25$

Thus there is no significant difference between the fourth and fifth iteration values.

Hence $u_1 = 1208,\ u_2 = 792,\ u_3 = 1042,\ u_4 = 458$

6.8 Hyperbolic Equations

The wave equation

$$\frac{\partial^2 u}{\partial t^2} = c^2 \frac{\partial^2 u}{\partial x^2}$$

is the most straightforward example of a hyperbolic partial differential equation. Its solution is the displacement function $u(x, t)$, defined for x values ranging from 0 to 1 and t values ranging from 0 to ∞, meeting the conditions at the start and at the end. As with parabolic equations, the solution advances in an open-ended region. However, in the case of hyperbolic equations, there are two initial conditions and two boundary conditions.

Convective type problems with vibrations, wave mechanics, and gas dynamics give rise to such equations.

6.9 Exercises

1. What is the classification of the equation $f_{xx} + 2f_{xy} + f_{yy} = 0$

2. Determine whether the following equation is elliptic or hyperbolic?

$$(x+1)u_{xx} - 2(x+2)u_{xy} + (x+3)u_{yy} = 0$$

3. Classify the equation

a) $y^2 u_{xx} - 2xyu_{xy} + x^2 u_{yy} + 2u_x - 3u = 0$

b) $x^2 \frac{\partial^2 u}{\partial x^2} + y^2 \frac{\partial^2 u}{\partial y^2} = x\frac{\partial u}{\partial x} - y\frac{\partial u}{\partial y}$

c) $3\frac{\partial^2 u}{\partial x^2} + 4\frac{\partial^2 u}{\partial x \partial y} + 6\frac{\partial^2 u}{\partial y^2} - 2\frac{\partial u}{\partial x} + \frac{\partial u}{\partial y} - u = 0$

4. In which parts of the (x, y) plane is the following equation elliptic?

$$\frac{\partial^2 u}{\partial x^2} + \frac{\partial^2 u}{\partial x \partial y} + (x^2 + 4y^2)\frac{\partial^2 u}{\partial y^2} = 2\sin(xy)$$

5. Solve the equation $u_{xx} + u_{yy} = 0$ for the square mesh with the boundary values as shown in Figure 6.7.

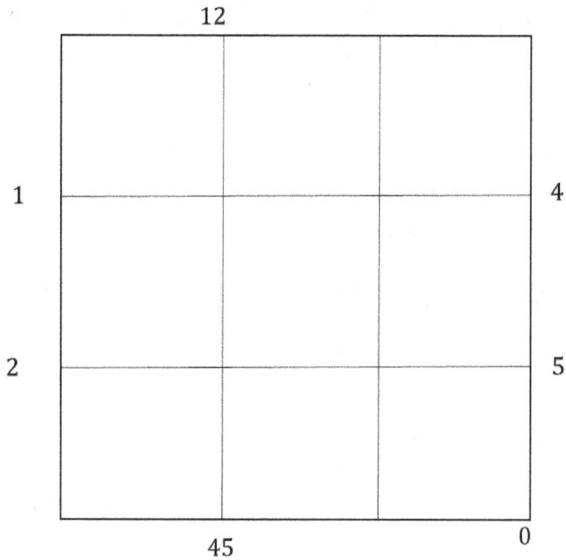

Figure 0.7 Square mesh for exercise 5

6. Solve the Laplace's equation $u_{xx} + u_{yy} = 0$ in the domain of Figure 6.8 by

(a) Jacobi's method,

(b) Gauss-Seidal method.

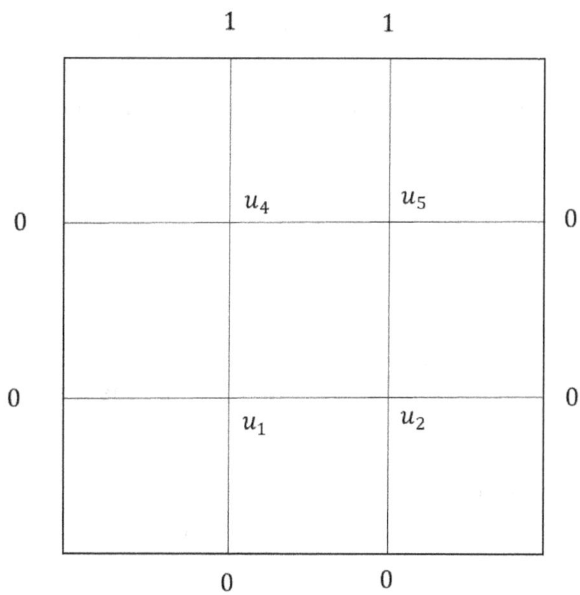

Figure 0.8 Square mesh for exercise 5

7. Apply explicit method to solve the diffusion equation
$u_t = u_{xx}, \qquad t > 0, 0 < x < 1;$
$u(0,t) = 0;$
$u(1,t) = 0;$
$u(x,0) = \sin \pi x.$
which has exact solution

$$u(x,y) = e^{-\pi^2 t} \sin \pi x$$

8. Repeat Problem 7, but this time take Δt 10% too large to satisfy the stability condition and attempt to compute solutions out to $t = 1$. Comment on what happens.

9. Apply Crank-Nicolson to the same PDE as in Problem 7. For each value of h, adjust the choice of Δt to obtain comparable accuracy in the pointwise norm to what was achieved above. Comment on your results.

10. Apply explicit method to solve the diffusion equation
$u_t = u_{xx}, \qquad t > 0, 0 < x < 1;$
$u(0,t) = 0;$
$u(1,t) = e^{-\pi^2 t/4};$
$u(x,0) = \sin \pi x/2.$
which has exact solution

$$u(x,y) = e^{-\pi^2 t/4} \sin \pi x/2$$

11. Find the solutions that depend only on r of the equation $u_{xx} + u_{yy} + u_{zz} = k^2 y$, where k is a positive constant. (Hint: Substitute $u = v/r$.)

12. Solve

$$\frac{\partial u}{\partial x} = \frac{\partial^2 u}{\partial^2 x^2}$$

in $0 < x < 5, t \geq 0$ given that $u(x,0) = 20$, $u(0,t) = 0, u(5,t) = 100$. Compute u for the time-step with $h = 1$ by the Crank-Nicholson method.

13. Find the solution of the parabolic equation $u_{xx} = 2u_t$, when $u(0,t) = u(4,t) = 0$ and $u(x,0) = x(4 - x)$, taking $h = 1$. Find the values up to $t = 5$.

14. Solve the heat equation

$$\frac{\partial u}{\partial t} = \frac{\partial^2 u}{\partial x^2}$$

subject to the conditions $u(0, t) = u(1, t) = 0$ and

$$u(x, 0) = \begin{cases} 2x, & 0 \le x \le \frac{1}{2} \\ 2(1 - x), & \frac{1}{2} \le x \le 1 \end{cases}$$

Take $h = 1/4$.

CHAPTER 7

Curve Fitting

7.1 Introduction to Curve Fitting and its Application

Data is frequently provided for specific values within a continuous range. Nevertheless, it may be necessary for us to obtain approximations at intervals that lie between these different values. The present section of this book outlines methodologies for fitting curves to such data in order to derive intermediate estimates. It could additionally require a simpler version of a function that is more involved. An approach to accomplish this is by calculating the function's values at many specific points within the desired range. After that, a more straightforward function can be constructed to accommodate these variables. Both of these applications are known to as curve fitting. An engineering application of experimental curve fitting is hypothesis testing. Fitted curves can also serve as a tool for data visualisation, allowing us to estimate values of a function in the absence of data and provide a summary of the relationships between two or more variables.

Here, we address the issue of approximating a function that is not well-known and can only be determined empirically at specific points. These values are prone to inherent errors, which can be particularly significant in various engineering and scientific applications. There are two main approaches to curve fitting that differ in terms of the error associated with the data. An approach uses the method of least square approximation to perform s curve fitting. This technique allows for the generation of a smooth approximation using noisy function values. This smooth approximation can be utilised to achieve a more accurate estimation of the derivative compared to exact polynomial interpolation.

Secondly, in cases where the data is highly accurate, a common method is to employ curve fitting techniques to establish a curve or a set of curves that accurately intersect each data point. Data of this nature typically originates from tables. Values for the density of water or for the heat capacity of gases as a function of temperature can be found in various examples. The process of estimating values between well-known discrete sets is known as interpolation.

7.2 Criterion to Obtain Best Fitted Curve

Curve fitting is the process of constructing a curve or mathematical function that closely approximates a given series of data. Through the process of curve fitting, a mathematical model can be constructed to establish the functional relationship between the observed dataset and parameter values. It is highly effective for modelling some natural processes mathematically. The example below uses a straight line function

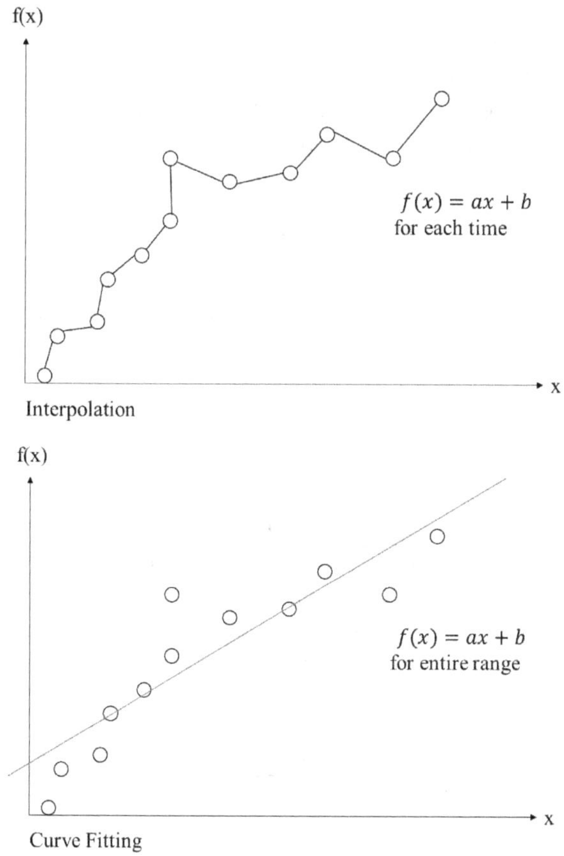

Figure 7.1 Two attempts to fit a "best" curve through data points

A straight line is described generically by $f(x) = ax + b$. The goal is to determine the coefficients 'a' and 'b' that result in a good fit of the function f(x) to the given data.

7.3 Method of Least Squares

Consider a collection of observed values $(x_1, f_1), (x_2, f_2), \dots, (x_n, f_n)$ and an approximation function $g(x)$. We calculate the sum of the squares of the differences between the observed values f_i and the estimated values $g(x_i)$, i.e.,

$$S = \sum_{i=1}^{n} \{f_i - g(x_i)\}^2 \tag{7.1}$$

The function $g(x)$ can have parameters α, β, γ. To determine these parameters, it is necessary to determine the required conditions for S to be minimised, which are as follows:

$$\frac{\partial S}{\partial \alpha} = 0, \frac{\partial S}{\partial \beta} = 0, \frac{\partial S}{\partial \gamma} = 0 \tag{7.2}$$

The normal equations are used to solve for the parameters of the best approximate function $g(x)$.

7.4 Curve Fitting by a Straight Line

The least squares method is used to find the line that best approximates the data by minimising the sum of the squared differences between the y-values on the line and the actual y-values. Therefore, the constants a_0 and a_1 need to be determined to minimise the least squares error. The general problem of fitting the best least squares line for a dataset $\{(x_i, y_i)\}_{i=1}^{m}$ involves minimizing the total error,

$$E = E_2(a_0, a_1) = \sum_{i=1}^{m} [y_i - (a_1 x_i + a_0)]^2 \tag{7.3}$$

with respect to the parameters a_0 and a_1. For a minimum to occur, we need both

$$\frac{\partial E}{\partial a_0} = 0 \text{ and } \frac{\partial E}{\partial a_1} = 0,$$

That is,

$$0 = \frac{\partial}{\partial a_0}\sum_{i=1}^{m}[y_i - (a_1x_i - a_0)]^2$$
$$= 2\sum_{i=1}^{m}[y_i - a_1x_i - a_0](-1)$$

And

$$0 = \frac{\partial}{\partial a_1}\sum_{i=1}^{m}[y_i - (a_1x_i + a_0)]^2$$
$$= 2\sum_{i=1}^{m}[y_i - a_1x_i - a_0](-x_i)$$

These equations can be simplified to the normal equations:

$$a_0.m + a_1\sum_{i=1}^{m}x_i = \sum_{i=1}^{m}y_i$$

And

$$a_0\sum_{i=1}^{m}x_i + a_1\sum_{i=1}^{m}{x_i}^2 = \sum_{i=1}^{m}x_i\,y_i$$

The solution to this system of equations is as follows,

$$a_0 = \frac{\sum_{i=1}^{m}{x_i}^2\sum_{i=1}^{m}y_i - \sum_{i=1}^{m}x_i\,y_i\sum_{i=1}^{m}x_i}{m\left(\sum_{i=1}^{m}{x_i}^2\right) - \left(\sum_{i=1}^{m}x_i\right)^2} \tag{7.4}$$

$$a_1 = \frac{m\sum_{i=1}^{m}x_i\,y_i - \sum_{i=1}^{m}x_i\sum_{i=1}^{m}y_i}{m\left(\sum_{i=1}^{m}{x_i}^2\right) - \left(\sum_{i=1}^{m}x_i\right)^2} \tag{7.5}$$

And then

$$a_0 = \bar{y} - a_1\bar{x} \tag{7.6}$$

Given that $\partial^2S/\partial\alpha_0^2$ and $\partial^2S/\partial\alpha_1^2$ are positive at points a_0 and a_1, it can be concluded that these values result in a minimum of S. In Eq. (7.6), $\bar{x}$ and $\bar{y}$ are the means of x and y, respectively. From Eq. (7.6), we have

$$\bar{y} = a_0 + a_1\bar{x}$$

Which shows the fitted straight line passes through the centroid of the data points.

A quality of fit is sometimes adopted. The correlation coefficient (cc) is a measure of the degree to which two variables change together.

$$cc = \sqrt{\frac{S_y - S}{S_y}} \tag{7.7}$$

Where

$$S_y = \sum_{i=1}^{m} (y_i - \bar{y})^2$$

(7.8)

and S is defined by,

$$S = [y_1 - f(x_1)]^2 + [y_2 - f(x_2)]^2 + \cdots + [y_m - f(x_m)]^2$$
$$= e_1^2 + e_2^2 + \cdots + e_m^2 \, , \tag{7.9}$$

The values of cc can range from -1 to 1. If it is close to 1, then the fit is considered to be good, although this is not always true.

7.4.1 Error Quantification of Linear Regression

Any line other than the one computed in Example 7.1 results in a larger sum of the squares of the residuals. The line is unique and, according to our selected criterion, is considered the most optimal line passing through the locations. Further characteristics of this fit can be revealed by thoroughly analysing the method used to calculate residuals. Remember that the sum of the squares is defined by Equation (7.9).

$$S_r = \sum_{i=1}^{n} e_i^2 = \sum_{i=1}^{n} (y_i - a_0 - a_1 x_i)^2 \tag{7.10}$$

In the previous situation, the square of the residual indicated the disparity between the data and a single estimate of the central tendency measure, which is the mean. The residual squared in Equation (7.10) represents the squared vertical distance between the data points and a different measure of central tendency, which is the straight line.

One can further extend the analogy to cases where the spread of points around the line is consistent across the entire range of the data and the distribution of these points about the line is normal. When these conditions are satisfied, least-squares regression offers the most accurate estimates of a_0 and a_1. Termed the maximum likelihood principle in statistics. If these conditions are satisfied, it is possible to calculate the standard deviation for the regression line as

$$S_{y/x} = \sqrt{\frac{S_r}{n-2}}$$

(7.11)

where $S_{y/x}$ is called the standard error of the estimate. The notation "y/x" indicates an error in the predicted value of y for a specific x value. It is important to observe that the division by $n-2$ is necessary due to the utilisation of two data-derived estimates (a_0 and a_1) in the calculation of S_r , resulting in the loss of two degrees of freedom. As with our discussion of the standard deviation, another justification for dividing by $n-2$ is that there is no such thing as the "spread of data" around a straight line connecting two points. Thus, for the case where $n = 2$, Eq. (7.11) yields a meaningless result of infinity. Just as was the case with the standard deviation, the standard error of the estimate quantifies the spread of the data. However, $S_{y/x}$ quantifies the spread around the regression line in contrast to the original standard deviation S_y that quantified the spread around the mean. The concepts mentioned can quantify the quality of our fit. This is especially valuable for comparing multiple regression models. We revisit the initial data to calculate the sum of squares around the mean for the dependent variable (y). This quantity is denoted as S_t. This represents the residual error linked to the dependent variable before regression. Upon completing the regression analysis, it is possible to calculate S_r, which represents the sum of the squared residuals in relation to the regression line. This describes the residual error left over following the regression. Hence, it is occasionally referred to as the unexplained sum of the squares. The discrepancy between two quantities, $S_t - S_r$ represents the enhancement or decrease in errors resulting from presenting the data as a linear model rather than an average. The scale-dependent magnitude of this quantity is normalised to S_t to yield the difference,

$$r^2 = \frac{S_t - S_r}{S_t} \tag{7.12}$$

where r^2 is called the coefficient of determination and r is the correlation coefficient.

For a perfect fit, when $S_r = 0$ and $r = r^2 = 1$, it indicates that the line accounts for 100% of the variability in the data. When $r = r^2 = 0$ and $S_t = S_r$, the fit represents no improvement. A more computationally convenient formulation for r is

$$r = \frac{n\sum x_i y_i - (\sum x_i)(\sum y_i)}{\sqrt{n\sum x_i^2 - (\sum x_i)^2}\ \sqrt{n - \sum y_i^2 - (\sum y_i)^2}} \tag{7.13}$$

Example 7.1 Fit a straight line to the x and y values in the first two columns of Table 7.1.

Solution. The following quantities can be computed:

$$n = 7 \qquad \sum x_i y_i = 119.5 \qquad \sum x_i^2 = 140$$

$$\sum x_i = 28 \qquad \bar{x} = \frac{28}{7} = 4$$

$$\sum y_i = 24 \qquad \bar{y} = \frac{24}{7} = 3.428571$$

Using Eqs. (7.4) and (7.5),

$$a_1 = \frac{7(119.5) - 28(24)}{7(140) - (28)^2} = 0.8392857$$

$$a_0 = 3.428571 - 0.8392857(4) = 0.07142857$$

Table 7.1 Computations for an error analysis of the linear fit

x_i	y_i	$(y_i - \bar{y})$	$(y_i - a_0 - a_1 x_i)^2$
1	0.5	8.5765	0.1687
2	2.5	0.8622	0.5625
3	2.0	2.0408	0.3473
4	4.0	0.3265	0.3265
5	3.5	0.0051	0.5896
6	6.0	6.6122	0.7972
7	5.5	4.2908	0.1993
$\sum$	24.0	22.7143	2.9911

Therefore, the least-squares fit is

$$y = 0.07142857 + 0.8392857x$$

Example 7.2 Find the straight line fitting the following data:

x_i	4	6	8	10	12
y_i	13.72	12.90	12.01	11.14	10.31

Solution. Let $y = a + bx$, be the straight line which fits the data. We have the normal equations

$$\frac{\partial S}{\partial a} = 0, \qquad \frac{\partial S}{\partial b} = 0$$

for determining a and b, where $S = \sum_{i=1}^{5} (y_i - a - bx_i)^2$

Thus, $\sum_{i=1}^{5} y_i - an - b\sum_{i=1}^{5} x_i = 0$

And, $\sum_{i=1}^{5} x_i y_i - a\sum_{i=1}^{5} x_i - b\sum_{i=1}^{5} {x_i}^2$

The coefficients are computed in the table below.

Table 7.2 The computation of coefficients for problem 7.2

	x_i	y_i	x_i^2	x_iy_i
	4	13.72	16	54.88
	6	12.90	36	77.40
	8	12.01	64	96.08
	10	11.14	100	111.40
	12	10.31	144	123.72
Sum	40	60.08	360	463.48

Thus, the normal equations are,

$$5a + 40b - 60.08 = 0$$
$$40a + 360b - 463.48 = 0$$

Solving these two equations we obtain,

$$a = 15.448, \qquad b = 0.429$$

Thus, $y = g(x) = 15.448 - 0.429\,x$, is the straight line fitting the data.

Example 7.3 Use the method of least square approximation to fit a straight line to the following observed data:

x_i	60	61	62	63	64
y_i	40	40	48	52	55

Solution. Let the straight line fitting the data be $y = a + bx$. The data values being large, we can use a change in variable by substituting $u = x - 62$ and $v = y - 48$.

Let $v = A + Bu$, be a straight line fitting the transformed data, where the normal equations for A and B are

$$\sum_{i=1}^{5} v_i = 5A + B\sum_{i=1}^{5} u_i$$
$$\sum_{i=1}^{5} u_iv_i = A\sum_{i=1}^{5} u_i + B\sum_{i=1}^{5} u_i^2$$

The computation of the various sums is given in the table below,

Table 7.3 The computation of coefficients for problem 7.3

x_i	y_i	u_i	v_i	u_iv_i	u_i^2
60	40	−2	−8	16	4
61	42	−1	−6	6	1
62	48	0	0	0	0
63	52	1	4	4	1
64	55	2	7	14	4
Sum		0	−3	40	10

Thus, the normal equations are,
$-3 = 5A$ and $40 = 10B$

$$\therefore A = -\frac{3}{5} + 4u$$

Or,
$$20u - 5y - 3 = 0$$

Transforming we get the line,
$$20\,(x - 62) - 5\,(y - 48) - 3 = 0$$

Or,
$$20\,x - 5y - 1003 = 0$$

Example 7.4 Find the least squares line approximating the following data.

x_i	1	2	3	4	5	6	7	8	9	10
y_i	1.3	3.5	4.2	5	7	8.8	10.1	12.5	13	15.6

Solution. We first extend the table to include x_i^2 and x_iy_i and sum the columns. This is shown in Table 7.4.

Table 7.4 The computation of coefficients for problem 7.4

x_i	y_i	x_i^2	x_iy_i	$P(x_i) = 1.538x_i - 0.360$
1	1.3	1	1.3	1.18
2	3.5	4	7.0	2.72

3	4.2	9	12.6	4.25
4	5	16	20	5.79
5	7	25	35	7,33
6	8.8	36	52.8	8.87
7	10.1	49	70.7	10.41
8	12.5	64	100	11.94
9	13	81	117	13.48
10	15.6	100	156	15.02
Sum 40	60.08	360	463.48	$E = \sum_{i=1}^{10}(y_i - P(x_i))^2 \approx 2.35$

The normal equations (7.4) and (7.5) imply that

$$a_0 = \frac{385(81) - 55(572.4)}{10(385) - (55)^2} = -0.360$$

$$a_1 = \frac{10(572.4) - 55(81)}{10(385) - (55)^2} = 1.538$$

so $P(x) = 1.538x - 0.360$. The graph of this line and the data points are shown in Figure 7-2. The approximate values given by the least squares technique at the data points are in Table 7.4.

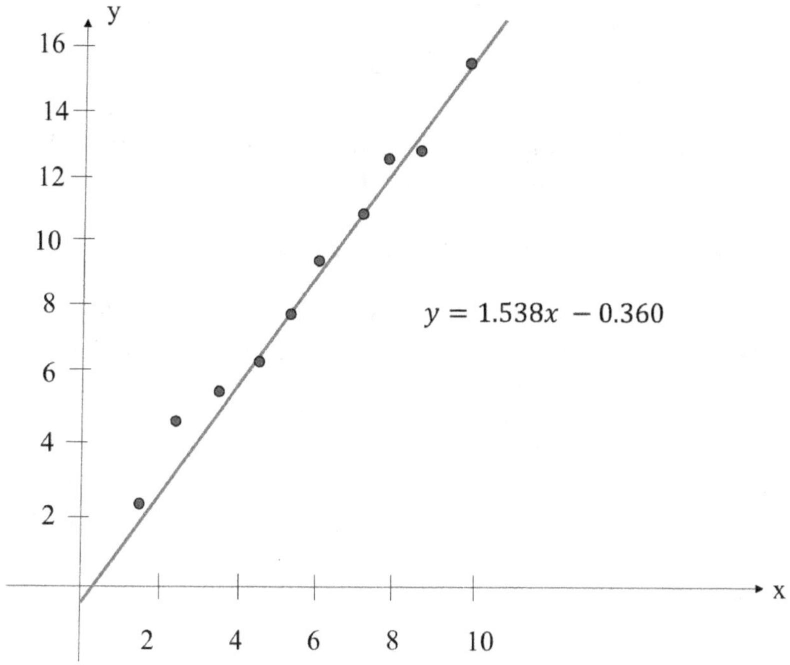

Figure 7.2 Line graph of the data points

Example 7.5 Compute the total standard deviation, the standard error of the estimate, and the correlation coefficient for the data in Example 7.1.

Solution. The summations are performed and presented in Table 17.1. The standard deviation is

$$S_y = \sqrt{\frac{22.7143}{7-1}} = 1.9457$$

and the standard error of the estimate is

$$S_{y/x} = \sqrt{\frac{2.9911}{7-2}} = 0.7735$$

Thus, because $S_{y/x} < S_y$,the linear regression model has merit. The extent of the improvement is quantified by

$$r^2 = \frac{22.7143 - 2.9911}{22.7143} - 0.868$$

Or, $$r = \sqrt{0.868} = 0.932$$

These results indicate that 86.8 percent of the original uncertainty has been explained by the linear model.

7.5 Linearization of Nonlinear Relationships

Linear regression is a powerful method for fitting a line to data. It is assumed that the relationship between the dependent and independent variables is linear. It is not always the case that the first step in any regression analysis is to plot and visually inspect the data to determine if a linear model is appropriate. As illustrated in Figure 7-3, the data exhibits a clear curvilinear pattern. Polynomial regression is suitable in certain instances. Transformations can be utilised by some to represent data in a manner suitable for linear regression analysis.

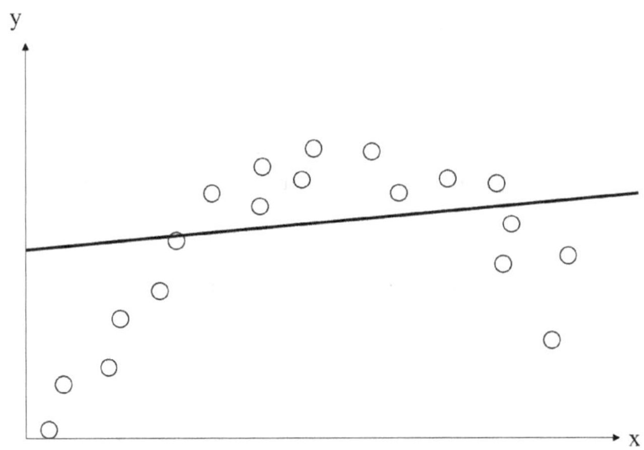

(a)

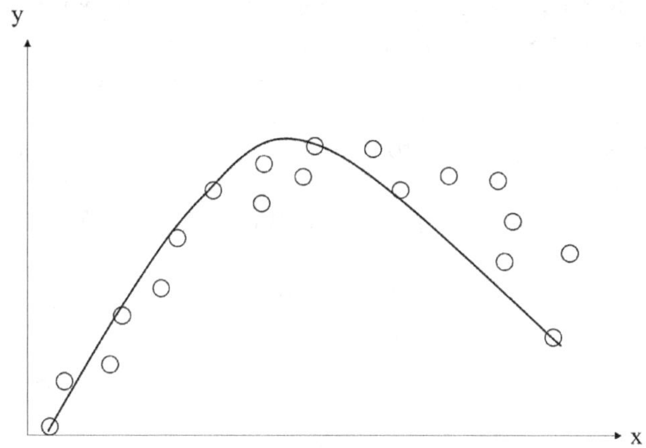

(b)

Figure 7.3 (a) Data that are ill-suited for linear least-squares regression, and (b) Indication that a parabola is preferable

One example is the exponential model

$$y = \propto_1 e^{\beta_1 x} \tag{7.14}$$

here $\propto_1$ and β_1 are constants. This model is commonly applied in various engineering disciplines to describe quantities that either increase (positive β_1) or decrease (negative β_1) in a manner directly proportional to their own size. For instance, phenomena like population growth or radioactive decay can demonstrate such characteristics. Another illustration of a nonlinear model is the basic power equation

$$y = \propto_2 x^{\beta_2} \tag{7.15}$$

where $\propto_2$ and β_2 are constant coefficients. This model shows broad applicability across different fields of engineering.

Another use of a nonlinear model is the saturation-growth-rate equation

$$y = \propto_3 \frac{x}{\beta_3 + x} \tag{7.16}$$

where $\propto_3$ and β_3 are constant coefficients. Nonlinear regression methods can be used to fit these equations to experimental data. Alternatively, one can utilise mathematical manipulations to convert the equations into a linear form. Next, one can use simple linear regression to fit the equations to the data.

One can linearize Eq. (7.14) by taking the natural logarithm of it

$$\ln y = \ln \propto_1 + \beta_1 x \ln e$$

But because $\ln e = 1$,

$$\ln y = \ln \propto_1 + \beta_1 x \qquad (7.17)$$

Plotting $\ln y$ against x results in a linear relationship with a slope of β_1 and an intercept of $\ln \propto_1$.

Equation (7.15) is linearized through the application of the base-10 logarithm, resulting in

$$\log y = \beta_2 \log x + \log \propto_2$$

Therefore, plotting $\log y$ against the $\log x$ will result in a linear relationship with a slope of β_2 and an intercept of $\log \propto_2$.

One can linearize Equation (7.14) by inverting it.1

$$\frac{1}{y} = \frac{\beta_3}{\propto_3}\frac{1}{x} + \frac{1}{\propto_3} \qquad (7.18)$$

Thus, a plot of $\frac{1}{y}$ versus $\frac{1}{x}$ will be linear, with a slope of $\frac{\beta_3}{\propto_3}$ and an intercept of $\frac{1}{\propto_3}$.

After transformation, the models can utilise linear regression to assess the constant coefficients. The specimens could be reverted to their initial state and utilised for predictive objectives. Example 17.4 illustrates this procedure for Eq. (7.15).

Example 7.6 Fit Eq. (7.15) to the data in Table 7.5 using a logarithmic transformation of the data.

Solution. Figure 7-4a is a plot of the original data in its untransformed state. Figure 7-4b shows the plot of the transformed data. A linear regression of the log-transformed data yields the result $\log y = 1.75 \log x - 0.3$

Table 7.5 Data to be fit to the power equation

x	y	$\log x$	$\log y$
1	0.5	0	−0.301
2	1.7	0.301	0.226
3	3.4	0.477	0.534
4	5.7	0.602	0.753
5	8.4	0.699	0.922

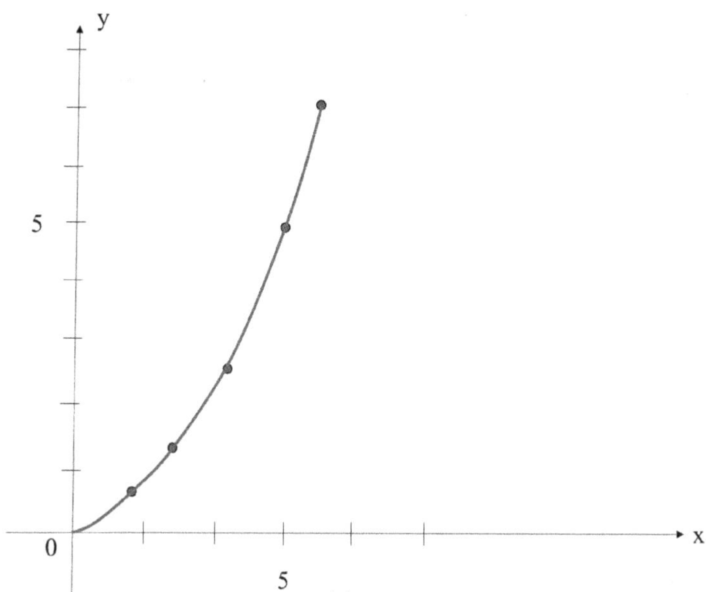

(a)

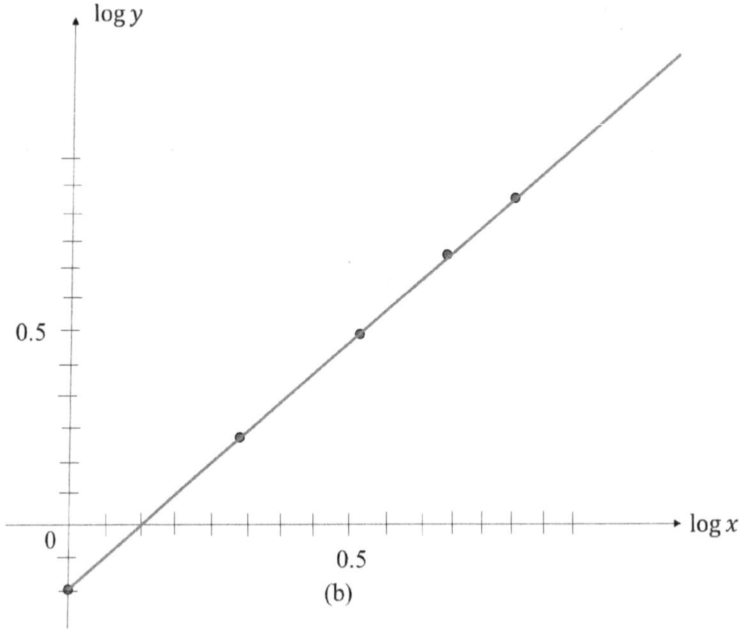

Figure 7.4 (a) Plot of untransformed data with the power equation that fits these data, and
(b) Plot of transformed data used to determine the coefficients of the power equation

Thus, the intercept, $\log \propto_2 = 0.300$, and therefore, by taking the antilogarithm,

$\propto_2 = 10^{-0.3} = 0.5$. The slope is $\beta_2 = 1.75$.

Consequently, the power equation is $y = 0.5x^{1.75}$

This curve, as plotted in Fig. 7-4a, indicates a good fit.

Example 7.7 Using the method of least squares, find constants a and b such that the function $y = ae^{bx}$ fits the following data: $(1.0, 2.473), (3.0, 6.722), (5.0, 18.274), (7.0, 49.673), (9.0, 135.026)$.

Solution. We have, $y = ae^{bx}$

Therefore, $\ln y =, \ln a + bx$

$$\Rightarrow Y = A_0 + A_1X$$

Where $Y = \ln y$, $A_0 = \ln a$, $A_1 = b$ and $X = x$.
The table of values is given below

Table 7.6 The computation of coefficients for problem 7.7

X	$Y = \ln y$	X^2	XY
1	0.905	1	0.905
3	1.905	9	5.715
5	2.905	25	14.525
7	3.905	49	27.335
9	4.905	81	44.145
25	14.525	165	92.625

We obtain, $\bar{X} = 5, A_0 = \bar{Y} = 2.905$

$$A_1 = \frac{5(92.625) - 25(14.525)}{5(165) - 625} = 0.5 = b$$

Then, $A_0 = \bar{Y} - A_1\bar{X} = 2.905 - 0.5(5) = 0.405$

Hence, $a = e^{A_0} = e^{0.405} = 1.499$

It follows that the required curve is of the form

$$y = 1.499e^{0.5x}$$

Example 7.8 Using the method of least squares, fit a curve of the form $y = \frac{x}{a+bx}$ to the following data:

$$(3, 7.148), (5.0, 10.231), (8.0, 13.509), (12, 16.434).$$

Solution. We have

$$y = \frac{x}{a + bx}$$

$$\Rightarrow \frac{1}{y} = \frac{a+bx}{x} = b + \frac{a}{x}$$
$$\Rightarrow Y = A_0 + A_1 X$$

Where $A_0 = b$, $A_1 = a$, $X = \frac{1}{x}$ and $Y = \frac{1}{y}$

The table of values is

Table 7.7 The computation of coefficients for problem 7.8

X	Y	X^2	XY
0.333	0.140	0.111	0.047
0.2	0.098	0.040	0.020
0.125	0.074	0.016	0.009
0.083	0.061	0.007	0.005
0.741	0.373	0.174	0.081

We obtain

$$\bar{X} = 0.185, \qquad \bar{Y} = 0.093$$
$$A_0 = b = \bar{Y} - a\bar{X} = 0.0331$$

$$A_1 = \frac{4(0.081) - 0.741(0.373)}{4(0.174) - (0.741)^2} = 0.324$$

Hence the required fit is $Y = o.0331 + 0.324(X)$, which simplifies to

$$y = \frac{x}{0.324 + 0.0331}$$

7.6 Polynomial Least Squares

A method has been developed to determine the equation of a straight line based on the least-squares criterion. Some engineering data, despite displaying a distinct pattern as illustrated in Figure 7-3, does not fit well with a linear model. In these instances, a curve would be more appropriate for fitting the data. In the previous section, a method to achieve this goal is through the use of transformations. One option is to fit polynomials to the data through polynomial regression.

Extending the least-squares procedure to fit data to a higher-order polynomial is feasible. For instance, consider fitting a second-degree polynomial or quadratic equation:

$$y = a_0 + a_1 x + a_2 x^2 + e$$

For this case the sum of the squares of the residuals is [compare with Eq. (7.10)]

$$S_r = \sum_{i=1}^{n} (y_i - a_0 - a_1 x_i - a_2 {x_i}^2)^2 \tag{7.19}$$

Applying the same method as described in the previous section, we calculate the derivative of Equation (7.19) with respect to each of the unidentified coefficients of the polynomial.

$$\frac{\partial S_r}{\partial a_0} = -2 \sum (y_i - a_0 - a_1 x_i - a_2 {x_i}^2)$$
$$\Rightarrow \frac{\partial S_r}{\partial a_1} = -2 \sum x_i (y_i - a_0 - a_1 x_i - a_2 {x_i}^2)$$
$$\Rightarrow \frac{\partial S_r}{\partial a_2} = -2 \sum {x_i}^2 (y_i - a_0 - a_1 x_i - a_2 {x_i}^2)$$

By setting these equations equal to zero and rearranging them, we can derive the following set of normal equations:

$$\begin{aligned} (n)a_0 + \left(\sum x_i\right) a_1 + \left(\sum {x_i}^2\right) a_2 &= \sum y_i \\ \left(\sum x_i\right) a_0 + \left(\sum {x_i}^2\right) a_1 + \left(\sum {x_i}^3\right) a_2 &= \sum x_i y_i \\ \left(\sum {x_i}^2\right) a_0 + \left(\sum {x_i}^3\right) a_1 + \left(\sum {x_i}^4\right) a_2 &= \sum {x_i}^2 y_i \end{aligned} \tag{7.20}$$

where all summations are from $i = 1$ through n.

It is important to observe that the three equations mentioned above are linear and involve three unknown variables: a_0, a_1, and a2. The unknown coefficients can be computed directly from the observed data.

In this situation, we observe that the task of finding a second-order polynomial that minimises the sum of squared errors is the same as solving a system of three linear equations that are all solved at the same time.

The two-dimensional scenario can be readily expanded to an mth-order polynomial:

$$y = a_0 + a_1x + a_2x^2 + \cdots + a_mx^m + e$$

The previous study can be easily expanded to this broader situation. Hence, it can be observed that finding the coefficients of a polynomial of order m is equivalent to solving a system of $m + 1$ simultaneous linear equation. In this particular example, the standard error is defined as

$$S_{y/x} = \sqrt{\frac{S_r}{n - (m + 1)}} \tag{7.21}$$

The quantity is split by $n - (m + 1)$ since $(m + 1)$ data-derived coefficients $(a_0, a_1, ..., a_m)$ were employed to compute S_r, resulting in a loss of $m - 1$ degrees of freedom. Furthermore, with the standard error, a coefficient of determination can be calculated for polynomial regression using Equation (7.12).

Example 7.9 Fit a second-order polynomial to the data in the first two columns of Table 7.8.

Solution. From the given data,

$$m = 2 \qquad \sum x_i = 15 \qquad \sum x_i^4 = 979$$

$$n = 6 \qquad \sum y_i = 152.6 \qquad \sum x_iy_i = 585.6$$

$$\bar{x} = 0.185 \qquad \sum x_i^2 = 55 \qquad \sum x_i^2y_i = 2488.8$$

$$\bar{y} = 0.093 \qquad \sum x_i^3 = 225$$

Table 7.8 The computation of coefficients for problem 7.9

x_i	y_i	$(y_i - \bar{y})^2$	$(y_i - a_0 - a_1 x_i - a_2 {x_i}^2)^2$
0	2.1	544.44	0.14332
1	7.7	314.47	1.00286
2	13.6	140.03	1.08158
3	27.2	3.12	0.80491
4	40.9	239.22	0.61951
5	61.1	1272.11	0.09439
$\sum$	152.6	2513.39	3.74657

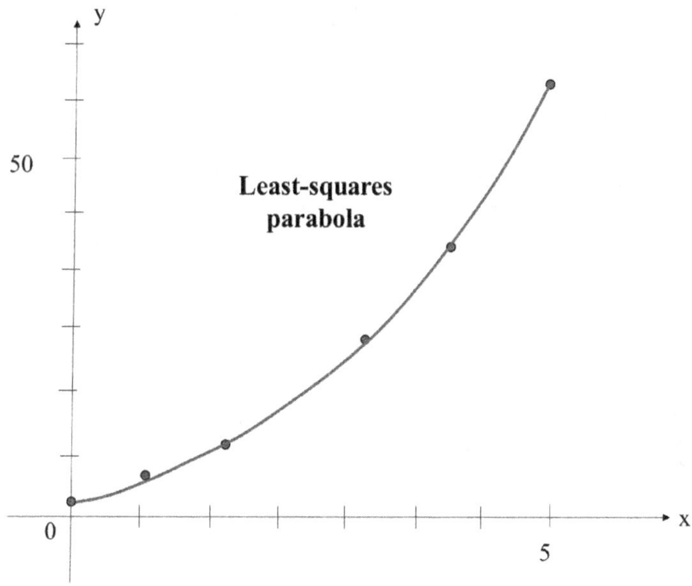

Figure 7.5 Fit of a second-order polynomial

Therefore, the simultaneous linear equations are

$$\begin{bmatrix} 6 & 15 & 55 \\ 15 & 55 & 225 \\ 55 & 225 & 979 \end{bmatrix} \begin{Bmatrix} a_0 \\ a_1 \\ a_2 \end{Bmatrix} = \begin{Bmatrix} 152.6 \\ 585.6 \\ 2488.8 \end{Bmatrix}$$

Solving these equations through a technique such as Gauss elimination gives

$a_0 = 2.47857\,, a_1 = 2.35929\,, a_2 = 1.86071.$

Therefore, the least-squares quadratic equation for this case is

$$y = 2.4785 + 2.35929x + 1.86071x^2$$

The standard error of the estimate based on the regression polynomial is [Eq. (7.21)]

$$S_{y/x} = \sqrt{\frac{3.74657}{5-3}} = 1.12$$

The coefficient of determination is

$$r^2 = \frac{2513.39 - 3.74657}{2513.39} = 0.99851$$

and the correlation coefficient is $r = 0.99925$.

The results suggest that the model has accounted for 99.851 percent of the initial uncertainty. This outcome substantiates the inference that the quadratic equation is a highly accurate fit, as is also apparent from Figure 7.5.

Example 7.10 Fit a polynomial of the second degree to the data points (x, y) given by $(0,1)$, $(1,6)$ and $(2,17)$.
Solution. For $n = 2$, Eq. (7.20) requires $\sum x_i$, $\sum x_i^2$, $\sum x_i^3$, $\sum x_i^4$, $\sum y_i$, $\sum x_i y_i$ and $\sum x_i^2 y_i$.

The table of values is as follows:

Table 7.9 The computation of coefficients for problem 7.10

x	y	x^2	x^3	x^4	xy	x^2y
0	1	0	0	0	0	0
1	6	1	1	1	6	6
2	17	4	8	16	34	68
3	24	5	9	17	40	74

The normal equations are

$$
\begin{aligned}
3a_0 + 3a_1 + 5a_2 &= 24 \\
3a_0 + 5a_1 + 9a_2 &= 40 \\
5a_0 + 9a_1 + 17a_2 &= 74
\end{aligned}
$$

Solving the system, we obtain,

$$a_0 = 1\,, a_1 = 2\,, a_2 = 3$$

The required polynomial is given by $Y = 1 + 2x + 3x^2$, and it can be seen that this fitting is exact.

Example 7.11 Fit a second degree parabola $y = a_0 + a_1x + a_2x^2$ to the data points (x_i, y_i) given

$$(1,0.63), (3, 2.05), (4, 4.08), (6, 10.78).$$

Solution. The table of values is as follows:

Table 7.10 The computation of coefficients for problem 7.11

x	y	x^2	x^3	x^4	xy	x^2y
1	0.63	1	1	1	0.63	0.63
3	2.05	9	27	81	6.15	18.45
4	4.08	16	64	256	16.32	65.28
6	10.78	36	216	1296	64.68	388.08
14	17.54	62	308	1634	87.78	472.44

The normal equations are

$$
\begin{aligned}
4a_0 + 14a_1 + 62a_2 &= 17.54 \\
14a_0 + 62a_1 + 30a_2 &= 87.78 \\
62a_0 + 30a_1 + 163a_2 &= 472.44
\end{aligned}
$$

Solving the system, we obtain,

$$a_0 = 1.24\,, a_1 = -1.05\,, a_2 = 0.44$$

7.7 Other nonlinear fits (exponential)

A common problem in the fields of engineering and physics involves fitting a sum of exponentials

$$y = f(x) = A_1 e^{\lambda_1 x} + A_2 e^{\lambda_2 x} + \cdots + A_n e^{\lambda_n x} \tag{7.22}$$

To a set of data points (x_i, y_i) , $i = 1,2, \dots, m$ where m is much greater than $2n$.

Here, we describe a computational method introduced by Moore [1974]. To simplify the presentation, we assume $n = 2$.

Then the function

$$y = A_1 e^{\lambda_1 x} + A_2 e^{\lambda_2 x} \tag{7.23}$$

is to be fitted to the data (x_i, y_i) , $i = 1,2, \dots, m$ where m is much greater than 4 . It is known that $y(x)$ satisfies a differential equation of the form

$$\frac{d^2y}{dx^2} = a_1 \frac{dy}{dx} + a_2 y \tag{7.24}$$

Where a_1 and a_2 have to be determined. Integrating Eq. (7.24), we get,

$$y'(x) - y'(x_0) = a_1[y(x) - y(0)] + a_2 \int_{x_0}^{x} y(x)dx \ , \tag{7.25}$$

Where x_0 is the initial value of x and $y'(x) = \frac{dy}{dx}$. By integrating Eq. (7.25), we get

$$y(x) - y(0) - (x - x_0)y'(x_0) = a_1 \int_{x_0}^{x} y(x)dx - a_1(x - x_0)y(x_0) + a_2 \iint_{x_0}^{x} y(x)dx\,dx \tag{7.26}$$

Now,

$$\iint_{x_0}^{x} y(x)dx\,dx = \int_{x_0}^{x} (x-t)y(t)dt$$

Hence, Eq. (7.26) becomes

$$\begin{aligned} y(x) - y(0) - (x - x_0)y'(x_0) \\ = a_1 \int_{x_0}^{x} y(x)dx - a_1(x - x_0)y(x_0) \\ + a_2 \int_{x_0}^{x} (x-t)y(t)dt \end{aligned} \tag{7.27}$$

Eq. (7.27) eliminates $y'(x_0)$ as follows. Consider x_1 and x_2 be two data points such that

$$x_0 - x_1 = x_2 - x_0 \tag{7.28}$$

Then Eq. (7.27) gives

$$\begin{aligned} y(x_1) - y(x_0) - (x_1 - x_0)y'(x_0) \\ = a_1 \int_{x_0}^{x_1} y(x)dx - a_1(x_1 - x_0)y(x_0) \\ + a_2 \int_{x_0}^{x_1} (x_1-t)y(t)dt \end{aligned} \tag{7.29}$$

And

$$\begin{aligned} y(x_2) - y(x_0) - (x_2 - x_0)y'(x_0) \\ = a_1 \int_{x_0}^{x_2} y(x)dx - a_1(x_2 - x_0)y(x_0) \\ + a_2 \int_{x_0}^{x_2} (x_2-t)y(t)dt \end{aligned} \tag{7.30}$$

Adding (7.29) and (7.30) and using Eq. (7.28), we get

$$y(x_1) + y(x_2) - 2y(x_0)$$
$$= a_1 \left[\int_{x_0}^{x_1} y(x)dx + \int_{x_0}^{x_2} y(x)dx \right]$$
$$+ a_2 \left[\int_{x_0}^{x_1} (x_1 - t)y(t)dt + \int_{x_0}^{x_2} (x_2 - t)y(t)dt \right] \tag{7.31}$$

Eq. (7.31) can now be used to set up a linear system of equations for a_1 and a_2 , and then we obtain λ_1 and λ_2 from the characteristic equation

$$\lambda^2 = a_1\lambda + a_2 \tag{7.32}$$

Finally, A_1 and A_2 can be obtained by the method of least squares or by the method of averages.

Example 7.12 Calculate Fitting exponential equation ($y = ae^{bx}$) Curve fitting using Least square method.

X	Y
0	0.10
0.5	0.45
1	2.15
1.5	9.15
2	40.35
2.5	180.75

Solution. The curve to be fitted is $y = ae^{bx}$

Taking logarithm on both sides, we get

$$\log_{10} v = \log_{10} a + bx \log_{10} e$$

$Y = A + Bx$ where $Y = \log_{10} v$, $A = \log_{10} a$, $B = b \log_{10} e$ which linear in Y, x

So, the corresponding normal equations are

$$\sum Y = nA + B\sum x$$
$$x\sum Y = A\sum x + B\sum x^2$$

The values are calculated using the following table,

Table 7.11 The computation of coefficients for problem 7.12

x	y	$Y = \log_{10} v$	x^2	$x.Y$
0	0.10	-1	0	0
0.5	0.45	-0.3468	0.25	-0.1734
1	2.15	0.3324	1	0.3324
1.5	9.15	0.9614	2.25	1.4421
2	40.35	1.6058	4	3.2117
2.5	180.75	2.2571	6.25	5.6427
$\sum x = 7.5$	$\sum y = 232.95$	$\sum Y = 232.95$	$\sum x^2 = 13.75$	$\sum x.Y = 10.4556$

Substituting these values in the normal equations

$$6A + 7.5B = 3.81$$
$$7.5A + 13.75B = 10.4556$$

Solving these two equations using elimination method,

we obtain $A = -0.9916,\ \ B = 1.3013$

$$a = anti \log_{10} A = anti \log_{10}(-0.9916) = 0.102$$

And

$$b = \frac{B}{\log_{10} e} = \frac{1.3013}{0.4343} = 2.9963$$

Now substituting these values in the equation is $y = ae^{bx}$, we get

$$y = 0.102e^{2.9963x}$$

Example 7.13 Fit a function of the form

$$y = A_1 e^{\lambda_1 x} + A_2 e^{\lambda_2 x}$$

To the data defined by (x, y)
(1,1.54), (1.1,1.67), (1.2,1.81) , (1.3,1.97), (1.4,2.15), (1.5,2.35), (1.6,2.58), (1.7,2.83), (1.8,3.11)

Let $x_0 = 1.2$, $x_1 = 1$, $x_2 = 1.4$ Then Eq. (7.31) gives

$$0.07 = a_1 \left[\iint_1^{1.2} y(x)dx + \int_{1.2}^{1.4} y(x)dx \right] + a_2 \left[-\int_1^{1.2} (1-t)y(t)dt + \int_{1.2}^{1.4} (1.4-t)y(t)dt \right]$$

Evaluating the integrals and simplifying, the above equation becomes

$$1.81a_1 + 2.180a_2 = 2.10$$

Again choosing $x_1 = 1.4$, $x_0 = 1.6$ and $x_2 = 1.8$, and evaluating the integrals, we obtain,

$$2.88a_1 + 3.104a_2 = 3.00$$

By solving,

$a_1 = 0.03204$ and $a_2 = 0.9364$

Eq. (7.32) gives

$$\lambda^2 - 0.03204 - 0.9364 = 0$$

From which we get,

$$\lambda_1 = 0.988 = 0.99$$
$$\lambda_2 = -0.96$$

$\therefore A_1 = 0.499$ and $A_2 = 0.491$

The above data was constructed from the function $y = \cos hx$ so that
$A_1 = A_2 = 0.5$,
$\lambda_1 = 1.0, \lambda_2 = -1.0$.

7.8 Exercises

1. Find the straight line fitting the following data:

x_i	4	6	8	10	12
y_i	13.72	12.90	12.01	11.14	10.31

2. Draw the graphs of the function $f(x) = \cos x - x$, in the range $[0, S/2)$ and find the location of the root of the equation $f(x) = 0$.

3. Use the method of least squares to fit a straight line for the following data points:

x_i	−1	0	1	2	3	4	5	6
y_i	10	9	7	5	4	3	0	−1

4. Use least-squares regression to fit a straight line to

x_i	0	2	4	6	9	11	12	15	17	19
y_i	5	6	7	6	9	8	7	10	12	12

Along with the slope and intercept, compute the standard error of the estimate and the correlation coefficient. Plot the data and the regression line. Then repeat the problem, but regress x versus y, that is, switch the variables. Interpret your results.

5. Use least-squares regression to fit a straight line to

x_i	6	7	11	15	17	21	23	29
y_i	29	21	29	14	21	15	7	7

Along with the slope and the intercept, compute the standard error of the estimate and the correlation coefficient. Plot the data and the regression line. If someone made an additional measurement of $x = 10, y = 10$, would you suspect, based on a visual assessment and the standard error, that the measurement was valid or faulty? Justify your conclusion.

6. Using the same approach as was employed to derive Eqs. (7.17) and (7.18), derive the least-squares fit of the following model:

$$y = a_1 x + e$$

That is, determine the slope that results in the least-squares fit for a straight line with a zero intercept. Fit the following data with this model and display the result graphically:

x_i	2	4	6	7	10	11	14	17	20
y_i	1	2	5	2	8	7	6	9	12

7. Use least-squares regression to fit a straight line to

x_i	1	2	3	4	5	6	7	8	9
y_i	1	1.5	2	3	4	5	6	10	13

(a) Along with the slope and intercept, compute the standard error of the estimate and the correlation coefficient. Plot the data and the straight line. Assess the fit.
(b) Recompute (a), but use polynomial regression to fit a parabola to the data. Compare the results with those of (a).

8. Fit the following data with the power model ($y = ax^b$). Use the resulting power equation to predict y at $x = 9$:

x_i	2.5	3.5	5	6	7.5	10	12.5	15	17.5	20
y_i	13	11	8.5	8.2	7	6.2	5.2	4.8	12	4.3

9. The following data provides the number of trucks with a particular weight at each hour of the day on one of the busy US highways. The data is provided as (h_i, N_i), where h_i is the hour and N_i is the number of trucks:

(1,16), (2,12), (3,14), (4,8), (5,24), (6,92), (7,311), (8,243), (9,558), (10,644), (11,768),

(12,838), (13,911), (14,897), (15,853), (16,860), (17,853), (18,875), (19,673), (20,378), (21,207), (22,142), (23,108), (24,62).

The distribution follows a Gaussian model of the form:

$$N = a_1 e^{\frac{-(x-a_2)^2}{a_3}}$$

a) Find the parameters a_1, a_2, a_3 and that would produce the best fit to the data.

b) On the same plot draw the model prediction vs. the data.

c) Calculate the R^2 of the model.

d) Calculate the sum of squares of the difference between the model prediction and the data.

10. The yield strength of steel decreases with elevated temperatures. The following experimental data provides the value of the temperature in 1000 degrees Celsius and the corresponding reduction factor in the yield strength.

(0.05,1), (0.1,1), (0.2,0.95), (0.3,0.92), (0.4,0.82), (0.5,0.6), (0.6,0.4), (0.7,0.15), (0.8,0.15) (0.9,0.1), (1.0,0.1).

Find the parameters in the following model that would provide the best fit to the data:

$$R = a_1 + \frac{a_2}{1 + a_3 e^{(a_4 x + a_5)}}$$

Also, on the same plot draw the model prediction vs. the data. Based on the model, what is the reduction in the yield strength at temperatures of 150, 450, and 750 Degrees Celsius?

11. Explain the method of least squares to fit a straight line of the form $Y = a_0 + a_1 x$ to the data (x_i, y_i) :

x_i	1	2	3	4	5	6
y_i	2.4	3.1	3.5	4.2	5	6

12. Find the parameters a_0 and a_1 so that $Y = a_0 + a_1 x$ to the data (x_i, y_i) :

x_i	0	1	2	3	4
y_i	1	2.9	4.8	6.7	8.6

13. If the straight line $Y = a_0 + a_1 x$ is the best fit to the set of data points $(x_1, y_1), (x_2, y_2), \ldots, (x_3, y_3)$, show that

$$\begin{vmatrix} x & y & 1 \\ \sum x_i & \sum y_i & n \\ \sum x_i^2 & \sum x_i y_i & \sum x_i \end{vmatrix}, \qquad i = 1, 2, \ldots, n$$

14. Find the values of a, b, c so that $Y = a + bx + cx^2$ is the best fit to the data:

x_i	0	1	2	3	4
y_i	1	0	3	10	21

15. Fit a least squares parabola $Y = a + bx + cx^2$ to the data:

x_i	0	1	2	3	4	5	6
y_i	71	89	67	43	31	18	9

16. Fit a function of the form $y = A_1 e^{\lambda_1 x} + A_2 e^{\lambda_2 x}$ to the data defined by (x, y):

x_i	0.1	1.1	1.2	1.3	1.4	1.5	1.6	1.7	1.8
y_i	1.54	1.67	1.81	1.97	2.15	2.35	2.58	2.83	3.11

17. Fit a power model to the data:
(1,1.93), (1.1,1.61), (1.2,2.27), (1.3,3.19), (1.4,3.19), (1.5,3.71), (1.6,4.29), (1.7,4.95), (1.8,6.07), (1.9,7.48), (2,8.72), (2.1,9.34), (2.2,11.62).

18. Find the least squares polynomials of degrees 1, 2, and 3 for the data in the following table. Compute the error E in each case. Graph the data and the polynomials.

x_i	0	0.15	0.31	0.5	0.6	0.75
y_i	1	1.004	1.031	1.117	1.223	1.422

19. Find the least squares polynomials of degrees 1, 2, and 3 for the data in the following table. Compute the error E in each case. Graph the data and the polynomials.

x_i	1.0	1.1	1.3	1.5	1.9	2.1
y_i	1.84	1.96	2.21	2.45	2.94	3.18

20. Determine the normal equations if the cubic polynomial $y = a_0 + a_1 x + a_2 x^2 + a_3 x^3$ is fitted to the data (x_i, y_i) , $i = 1,2, \dots, m$.

21. Determine the constants a, b by the method of least squares, such that the curve $y = ae^{bx}$ fits the data:

x_i	2	4	6	8	10
y_i	4.077	11.084	30.128	81.897	222.62

22. Fit the following data with the power model $(y = ax^b)$.

x_i	61	26	7	2.6
y_i	350	400	500	600

CHAPTER 8

Polynomial Interpolation

8.1 Introduction

Interpolation involves deriving a simplified function from a set of discrete data points. This function accurately reproduces the given data points and can be used to estimate points between them. Let's consider an example where a function $f(x)$ is not explicitly defined, but we have its values at a finite number of points $\{x_i, i = 1, 2, \cdots, n\}$. The objective is to determine the value of f at a given point x, which falls within the range of x_j and x_k, where j and k are integers ranging from 1 to n. To obtain this, one can start by approximating function f using a known function. Then, the value of this approximation at point x can be determined. This process is known as interpolation. Polynomials are frequently utilised for interpolation due to their ease of evaluation, differentiation, and integration. This method is commonly referred to as polynomial interpolation. It can be proven that given $n + 1$ data points it is always possible to find a polynomial of order n to pass through or reproduce the $n + 1$ points.

In engineering and science, it is common to have a set of data points that are obtained through sampling or experimentation. These data points represent the values of a function for a limited range of the independent variable. It is frequently necessary to interpolate, or calculate the value of the function for an intermediate independent variable value. Interpolation is widely used in various disciplines such as numerical analysis, data engineering, computer graphics and mathematics. Interpolation offers several benefits which are as follows:

1. Value estimation: Interpolation can be utilised to estimate values between known data points, resulting in the creation of smooth curves and surfaces.
2. Data Compression: Interpolation can be used to reduce the amount of data that needs to be stored or transmitted by reducing the number of data points needed to represent a curve or surface.

3. Image and Video Processing: Interpolation is a commonly employed technique in picture and video processing to enhance image resolution or refine pixelated images.
4. Numerical Analysis: Interpolation is commonly employed in numerical analysis to solve differential equations by providing an approximation of the equation's solution at intermediate points.

Interpolation is a useful method for estimating values between known data points, which can help reduce the amount of data needed to represent a curve or surface, or analyse photos and videos.

8.2 Interpolation and Extrapolation

Interpolation and extrapolation are techniques commonly employed to approximate the value of a function at a specific point by utilising known values from nearby points. The following table shows the difference between extrapolation and interpolation.

Table 8.1 Difference between Interpolation and Extrapolation

Factors	Interpolation	Extrapolation
Describing	Interpolation is the process of estimating the value of a function at a point within the domain of the function based on known values at surrounding points.	Extrapolation is the process of estimating the value of a function outside the range of the known data.
Goal	The goal of interpolation is to approximate the unknown value of the function as closely as possible based on the known data.	In extrapolation, the goal is to extend the known data to predict the behavior of the function in a new and unknown region.

8.3 Lagrange Interpolation

The problem of finding a linear polynomial that goes through the points (x_0, y_0) and $(x_1,\ y_1)$ is equivalent to approximating a function f, where $f(x_0) = y_0$ and $f(x_1) = y_1$, using a first-degree polynomial that matches the values of f at those points. Applying this polynomial to approximate values within a specific interval is known as polynomial interpolation.

Let's consider the functions

$L_0(x) = \frac{x-x_1}{x_0-x_1}$ and $L_1(x) = \frac{x-x_0}{x_1-x_0}$

The linear Lagrange polynomial that interpolates between the points (x_0, y_0) and (x_1, y_1) is given by

$$P(x) = L_0(x)f(x_0) + L_1(x)f(x_1) = \frac{x-x_1}{x_0-x_1} f(x_0)\frac{x-x_0}{x_1-x_0} f(x_1) \tag{8.1}$$

It is noted that

$L_0(x_0) = 1\,,\ L_0(x_1) = 0\,,\ L_1(x_0) = 0$, and $L_1(x_1) = 1$

which implies that

$$P(x_0) = 1 \cdot f(x_0) + 0 \cdot f(x_1) = f(x_0) = y_0 \tag{8.2}$$

And

$$P(x_1) = 0 \cdot f(x_0) + 1 \cdot f(x_1) = f(x_1) = y_1 \tag{8.3}$$

Hence P is the unique polynomial of degree at most one that passes through (x_0, y_0) and (x_1, y_1).

Example 8.1 Determine the linear Lagrange interpolating polynomial that passes through the points $(2, 4)$ and $(5, 1)$.

Solution. In this case we have

$L_0(x) = \frac{x-5}{2-5} = -\frac{1}{3}(x-5)$ and $L_1(x) = \frac{x-2}{5-2} = \frac{1}{3}(x-2)$

So $\quad P(x) = -\frac{1}{3}(x-5) \cdot 4 + \frac{1}{3}(x-2) \cdot 1 = -\frac{4}{3}x + \frac{20}{3} + \frac{1}{3}x - \frac{2}{3} = -x + 6$

The graph of $y = P(x)$ is shown in Figure 8.1.

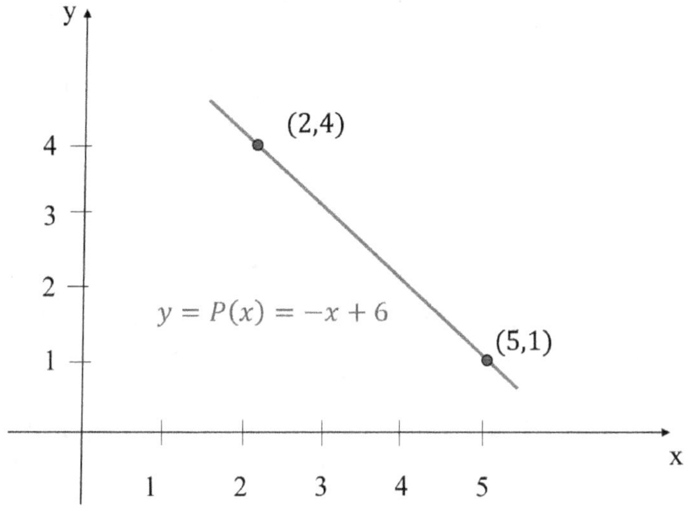

Figure 8.1 The graph of $y = P(x)$

Let's look at the construction of a polynomial with a maximum degree of n that passes through $n+1$ points in order to provide a broader understanding of linear interpolation

$$(x_0, f(x_0)), (x_1, f(x_1)), \ldots, (x_n, f(x_n)) \tag{8.4}$$

For this case, we start by creating a function $L_{n,k}(x)$ for each $k = 0, 1, \ldots, n$. These functions have the property that $L_{n,k}(x_i) = 0$ when $i \neq k$ and $L_{n,k}(x_k) = 1$. In order to fulfil the condition of $L_{n,k}(x_k) = 0$ for every $i \neq k$, it is necessary for the numerator of $L_{n,k}(x)$ to include the specific term

$$(x - x_0)(x - x_1) \cdots (x - x_{k-1})(x - x_{k+1}) \cdots (x - x_n) \tag{8.5}$$

To satisfy $L_{n,k}(x_k) = 1$, the denominator of $L_{n,k}(x)$ must be this same term but evaluated at $x = x_k$. Thus

$$L_{n,k}(x) = \frac{(x - x_0) \cdots (x - x_{k-1})(x - x_{k+1}) \cdots (x - x_n)}{(x_k - x_0) \cdots (x_k - x_{k-1})(x_k - x_{k+1}) \cdots (x_k - x_n)} \tag{8.6}$$

A sketch of the graph of a typical $L_{n,k}$ (when n is even) is shown in Figure 8.2.

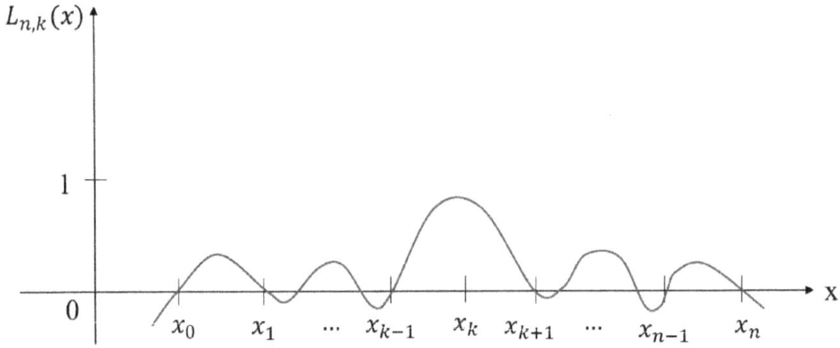

Figure 8.2 the graph of a typical $L_{n,k}$ (when n is even)

The description of the interpolating polynomial becomes clear once the form of $L_{n,k}$ is understood. The nth Lagrange interpolating polynomial is defined in the following theorem.

Theorem 8.1 Given a set of $n+1$ distinct numbers $x_0, x_1, \ldots x_n$ and a function f defined at these numbers, there exists a unique polynomial $P(x)$ of degree at most n that satisfies

$$f(x_k) = P(x_k), \qquad \text{for each } k = 0,1, \ldots, n$$

This polynomial is given by

$$P(x) = f(x_0)L_{n,0}(x) + \cdots + f(x_n)L_{n,n}(x) = \sum_{k=0}^{n} f(x_k)\, L_{n,k}(x) \tag{8.7}$$

where, for each $k = 0, 1, \ldots, n$,

$$L_{n,k}(x) = \frac{(x - x_0)(x - x_1)\cdots(x - x_{k-1})(x - x_{k+1})\cdots(x - x_n)}{(x_k - x_0)(x_k - x_1)\cdots(x_k - x_{k-1})(x_k - x_{k+1})\cdots(x_k - x_n)} \tag{8.8}$$

$$= \prod_{\substack{i=0 \\ i \neq k}}^{n} \frac{(x - x_i)}{(x_k - x_i)}.$$

When there is no uncertainty regarding the degree, $L_{n,k}(x)$ will be represented simply as $L_k(x)$.

Example 8.2 (a) Use the numbers (called nodes) $x_0 = 2,\ x_1 = 2.75$, and $x_2 = 4$ to find the second Lagrange interpolating polynomial for $f(x) = 1/x$.
(b) Use this polynomial to approximate $f(3) = 1/3$.

Solution. (a) We first determine the coefficient polynomials $L_0(x)$, $L_1(x)$, and $L_2(x)$. In nested form they are

$$L_0(x) = \frac{(\mathrm{x} - 2.75)(\mathrm{x} - 4)}{(2 - 2.5)(2 - 4)} = \frac{2}{3}(\mathrm{x} - 2.75)(\mathrm{x} - 4)$$
$$L_1(x) = \frac{(\mathrm{x} - 2)(\mathrm{x} - 4)}{(2.75 - 2)(2.75 - 4)} = \frac{16}{15}(\mathrm{x} - 2)(\mathrm{x} - 4)$$
$$L_0(x) = \frac{(\mathrm{x} - 2)(\mathrm{x} - 2.75)}{(4 - 2)(4 - 2.5)} = \frac{2}{5}(\mathrm{x} - 2)(\mathrm{x} - 2.75)$$

Also, $f(x_0) = f(2) = \frac{1}{2}, f(x_1) = f(2.75) = 4/11$, and $f(x_2) = f(4) = 1/4$, so

$$P(x) = \sum_{k=0}^{2} f(x_k)L_k(x)$$
$$= \frac{1}{3}(\mathrm{x} - 2.75)(\mathrm{x} - 4) - \frac{64}{165}((\mathrm{x} - 2)(\mathrm{x} - 4) + \frac{1}{10}(\mathrm{x} - 2)(\mathrm{x} - 2.75)$$
$$= \frac{1}{22}x^2 - \frac{35}{88}x + \frac{49}{44}$$

(b) An approximation to $f(3) = 1/3$ (see Figure 8-3) is

$$\mathrm{f}(3) \approx \mathrm{P}(3) = \frac{9}{22} - \frac{105}{88} + \frac{49}{44} = \frac{29}{88} \approx 0.32955$$

It is important to note that in the initial section of this chapter, it was determined that no Taylor polynomial enlarged around $x_0 = 1$ could be employed to reasonably simulate the function $f(x) = 1/x$ at $x = 3$.

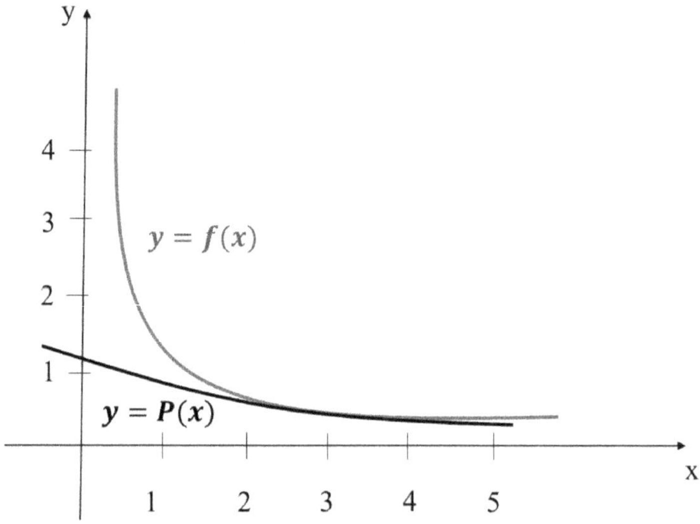

Figure 8.3 Approximation to $f(3) = 1/3$

Theorem 8.2 Consider a set of distinct numbers $x_0, x_1, \cdots, x_n$ within the interval $[a, b]$, and $f \not\approx C^{n+1}[a, b]$. For every x in the interval $[a, b]$, there exists an unidentified number $\xi(x)$ between $x_0, x_1, \ldots, x_n$, and so in (a, b), with

$$f(x) = P(x) + \frac{f^{(n+1)}(\xi(x))}{(n+1)!}(x - x_0)(x - x_1) \cdots (x - x_n), \tag{8.9}$$

where $P(x)$ is the interpolating polynomial given in Eq. (8.7).

Proof: Note first that if $x = x_k$, for any $k = 0, 1, \cdots, n$, then $f(x_k) = P(x_k)$, and choosing $\xi(x_k)$ arbitrarily in (a, b) yields Eq. (8.9).

If $x \neq x_k$, *for all* $k = 0, 1, \ldots, n$, define the function g for t in [a,b] by

$$g(t) = f(t) - P(t) - [f(x) - P(x)]\frac{(t - x_0)(t - x_1) \ldots (t - x_n)}{(x - x_0)(x - x_1) \ldots (x - x_n)}$$

$$= f(t) - P(t) - [f(x) - P(x)]\prod_{i=0}^{n}\frac{t - x_i}{x - x_i}.$$

As $f \in C^{n+1}[a, b]$ and $P \in C^{\infty}[a, b]$, it follows that $g \in C^{n+1}[a, b]$. For $t = x_k$, we have

$$g(x_k) = f(x_k) - P(x_k) - [f(x) - P(x)] \prod_{i=0}^{n} \frac{x_k - x_i}{x - x_i}$$
$$= 0 - [f(x) - P(x)] \cdot 0 = 0.$$

Moreover,

$$g(x) = f(x) - P(x) - [f(x) - P(x)] \prod_{i=0}^{n} \frac{x - x_i}{x - x_i}$$
$$= f(x) - P(x) - [f(x) - P(x)] = 0$$

Thus $g \in C^{n+1}[a, b]$, and g is zero at the $n +$ distinct numbers $x, x_0, x_1, \cdots, x_n$. By Generalized Rolle's Theorem, there exists a number ξ in (a, b) for which $g^{n+1}(\xi) = 0$.

$$\therefore 0 = g^{n+1}(\xi) = f^{n+1}(\xi) - P^{n+1}(\xi)$$
$$- [f(x) - P(x)] \frac{d^{n+1}}{dt^{n+1}} \left[\prod_{i=0}^{n} \frac{t - x_i}{x - x_i} \right]_{t=\xi} \tag{8.10}$$

The polynomial $P(x)$ has a maximum degree of n, hence the $(n + 1)$st derivative, $P^{(n+1)}(x)$, is equal to zero. Again, $\prod_{i=0}^{n} \frac{t-x_i}{x-x_i}$ is a polynomial of degree $(n + 1)$, so

$$\prod_{i=0}^{n} \frac{t - x_i}{x - x_i} = \left[\frac{1}{\prod_{i=0}^{n} (x - x_i)} \right] t^{n+1} + ((\text{lower} - \text{degree terms in t})$$

And

$$\frac{d^{n+1}}{dt^{n+1}} \left[\prod_{i=0}^{n} \frac{t - x_i}{x - x_i} \right] = \frac{(n + 1)!}{\prod_{i=0}^{n} (x - x_i)}$$

Equation (8.10) now becomes

$$0 = f^{n+1}(\xi) - 0 - [f(x) - P(x)] \frac{(n + 1)!}{\prod_{i=0}^{n} (x - x_i)},$$

and, upon solving for $f(x)$, we have

$$f(x) = P(x) + \frac{f^{n+1}(\xi)}{(n+1)!}\prod_{i=0}^{n}(x - x_i)$$

(8.11)

The error formula shown in Theorem 8.3 holds significant theoretical significance due to the widespread utilisation of Lagrange polynomials in the derivation of numerical differentiation and integration techniques. The Lagrange error formula is used to derive error boundaries for these strategies.

It is important to observe that the error configuration for the Lagrange polynomial closely resembles that of the Taylor polynomial. The nth Taylor polynomial pertaining to x_0 encompasses the entirety of the available information at x_0 and incorporates an error term in the following way.

$$\frac{f^{n+1}(\xi(x))}{(n+1)!}(x - x_0)^{n+1}$$

A Lagrange polynomial with a degree of n utilises data at the specific values $x_0, x_1, \cdots, x_n$. Instead of $(x - x_0)^n$, its error formula involves multiplying the $n + 1$ components $(x - x_0)(x - x_1) \cdots (x - x_n)$:

$$\frac{f^{n+1}(\xi(x))}{(n+1)!}(x - x_0)(x - x_1) \cdots (x - x_n)$$

(8.12)

Example 8.3 In Example 8.2 we found the second Lagrange polynomial for $f(x) = 1/x$ on $[2, 4]$ using the nodes $x_0 = 2,\ x_1 = 2.75$, and $x_2 = 4$. Determine the error form for this polynomial, and the maximum error when the polynomial is used to approximate $f(x)$ for $x \in [2, 4]$.

Solution. Because $f(x) = x - 1$, we have
$f'(x) = -x^2$, $f''(x) = 2x^{-3}$, and $f'''(x) = -6x^{-4}$

As a consequence, the second Lagrange polynomial has the error form

$$\frac{f'''(\xi(x))}{3!}(x-x_0)(x-x_1)(x-x_2)$$
$$= -(\xi(x))^{-4}(x-2)(x-2.75)(x-4)$$

for $x \in [2,4]$.

The maximum value of $(\xi(x))^{-4}$ on the interval is $2^{-4} = \frac{1}{16}$. We now need to determine the maximum value on this interval of the absolute value of the polynomial

$$g(x) = (x-2)(x-2.75)(x-4) = x^3 - \frac{35x^2}{4} + \frac{49x}{2} - 22$$

Because

$$D_x\left(x^3 - \frac{35x^2}{4} + \frac{49x}{2} - 22\right) = 3x^2 - \frac{35x}{4} + \frac{49}{2}$$
$$= \frac{1}{2}(3x-7)(2x-7)$$

the critical points occur at

$x = {}^{7}/_{3}$, with $g({}^{7}/_{3}) = {}^{25}/_{108}$, and $x = {}^{7}/_{2}$, with $g({}^{7}/_{2}) = -{}^{9}/_{16}$

Hence, the maximum error is

$$\frac{f'''(\xi(x))}{3!}|(x-x_0)(x-x_1)(x-x_2)| \le \frac{1}{16.6}\left|-\frac{9}{16}\right| = \frac{3}{512}$$
$$\approx 0.00586$$

The following example demonstrates how to make use of the error formula in generating a data table that ensures that the initial interpolation error remains within a specified bound.

Example 8.4 Assume that a table for the function $f(x) = e^x$, where x is a value in the interval $[0,1]$. It is assumed that the number of decimal places to be provided per entry is $d \ge 8$, and that the step size, which represents the difference between adjacent x-values, is h. What is the step size, denoted as h, that guarantees linear interpolation yields an absolute error of no more than 10^{-6} for all values of x inside the interval $[0,1]$?

Solution. Let $x_0, x_1, \cdots$ be the numbers at which f is evaluated, x be in $[0,1]$, and suppose j satisfies $x_j \le x \le x_{j+1}$. Eq. (8.9) implies that the error in linear interpolation is

$$|f(x) - P(x)| = \left|\frac{f^2(\xi)}{2!}(x - x_j)(x - x_{j+1})\right|$$
$$= \frac{|f^2(\xi)|}{2}|(x - x_j)||(x - x_{j+1})|$$

The step size is h, so $x_j = jh, x_{j+1} = (j + 1)h$, and

$$|f(x) - P(x)| \le \frac{|f^2(\xi)|}{2!}|(x - jh)(x - (j + 1)h)|.$$

Hence

$$|f(x) - P(x)| \le \frac{\max_{\xi \in [0,1]} e^{\xi}}{2} \max_{x_j \le x \le x_{j+1}} |(x - jh)(x - (j + 1)h)|$$
$$\le \frac{e}{2} \max_{x_j \le x \le x_{j+1}} |(x - jh)(x - (j + 1)h)|.$$

Consider the function $g(x) = (x - jh)(x - (j + 1)h)$, for $jh \le x \le (j + 1)h$. Because

$$g'(x) = (x - (j + 1)h) + (x - jh) = 2\left(x - jh - \frac{h}{2}\right),$$

the only critical point for g is at $x = jh + {}^{h}/_{2}$, with $g(jh + {}^{h}/_{2}) = (h/2)^2 = h^2/4$.

Since $g(jh) = 0$ and $g((j + 1)h) = 0$, the maximum value of $|g'(x)|$ in $[jh, (j + 1)h]$ must occur at the critical point which implies that

$$|f(x) - P(x)| \le \frac{e}{2} \max_{x_j \le x \le x_{j+1}} |g(x)| \le \frac{e}{2} \cdot \frac{h^2}{4} = e\frac{h^2}{8}.$$

Consequently, to ensure that the error in linear interpolation is bounded by 10^{-6}, it is sufficient for h to be chosen so that

$$e\frac{h^2}{8} \le 10^{-6}.$$

This implies that $h < 1.72 \times 10^{-3}$

Because $n = (1 - 0)/h$ must be an integer, a reasonable choice for the step size is $h = 0.001$.

Example 8.5 Using Lagrange's interpolation formula, find the form of the function $y(x)$ from the following table.

x	y
0	−12
1	0
3	12
4	24

Solution. Since $y = 0$ when $x = 1$, it follows that $x - 1$ is a factor. Let $y(x) = (x - 1)\,R(x)$.

Then $R(x) = {}^{y}/_{x-1}$

We now tabulate the values of x and $R(x)$.

Table 8.2 Tarbulated values of x and $R(x)$

x	$R(x)$
0	12
3	6
4	8

Applying Lagrange's formula to the above table, we find

$$R(x) = \frac{(x-3)(x-4)}{(-3)(-4)}(12) + \frac{(x-0)(x-4)}{(3)(3-4)}(6) + \frac{(x-0)(x-3)}{(4-0)(4-3)}(8)$$

$$= (x-3)(x-4) - 2x(x-4) + 2x(x-3)$$
$$= x^2 - 5x + 12$$

Hence the required polynomial approximation to $y(x)$ is given by

$$y(x) = (x-1)(x^2 - 5x + 12).$$

8.4 Newton Interpolation and Divided Differences

The application of the Lagrange form for the interpolating polynomial provides a highly organised framework, although it presents challenges in terms of practical calculation. One contributing factor is the necessity to do a complete recalculation of all $L_i^{(n)}$ functions when including a new point into the set of nodes. This is due to the inherent difficulty in expressing p_{n+1} in terms of p_n using the Lagrange architecture. This issue is avoided by using an alternative version of the polynomial called the Newton form, which makes it simple to write p_{n+1} in terms of p_n. The fundamental outcome is encompassed within the subsequent theorem.

Theorem 8.3 (Newton Interpolation Construction) Let p_n denote the polynomial that interpolates the function f at the nodes x_i , where $i = 0,1,2,3,\dots,$n. . Consider the polynomial p_{n+1}, which is used to interpolate the function f at the given nodes x_i,

$i = 0,1,2,3,\dots,n,n+1$. Let p_{n+1} be expressed as

$$p_{n+1}(x) = p_n(x) + a_{n+1} w_n(x) \tag{8.13}$$

Where

$$w_n(x) = \prod_{i=0}^{n} (x - x_i)$$

$$a_{n+1} = \frac{f(x_{n+1}) - p_n(x_{n+1})}{w_n(x_{n+1})} \tag{8.14}$$

And $p_0(x) \equiv a_0 = f(p_0)$.

Proof: To demonstrate that p_{n+1}, as stated in (8.13), meets the interpolation criteria, we can use the fact that the interpolation polynomial is unique. Given that $k < n$, we can deduce that $w_n(x_k) = 0$, so

$$p_{n+1}(x_k) = p_n(x_k) + a_{n+1}w_n(x_k) = p_n(x_k) = f(x_k)$$

hence, p_{n+1} interpolates all but the last point. To check x_{n+1}, we directly compute:

$$\begin{aligned} p_{n+1}(x_{n+1}) &= p_n(x_{n+1}) + a_{n+1}w_n(x_k) \\ &= p_n(x_{n+1}) + f(x_{n+1}) - p_n(x_{n+1}) \\ &= f(x_{n+1}) \end{aligned}$$

Therefore, p_{n+1} interpolates at every node. Additionally, it is evident that p_{n+1} is a polynomial with a degree less than or equal to $n+1$, thus concluding our task.

For $\{a_k\}$ and w_n as defined in Theorem 8.3, we have

$$\begin{aligned} p_n(x) &= a_0 + a_1(x - x_0) + a_2(x - x_0)(x - x_1) + \cdots \\ &+ a_n(x - x_0) \cdots (x - x_{n-1}) \\ &\sum_{k=0}^{n} a_k w_{k-1}(x) \end{aligned}$$

where we have used $w_{-1}(x) \equiv 1$.

Proof: The Newton form of the interpolating polynomial can be used to efficiently generate and evaluate interpolating polynomials, similar to Homer's formula for nested multiplication. We have

$$\begin{aligned} p_n(x) &= a_0 + a_1(x - x_0) + a_2(x - x_0)(x - x_1) + \cdots \\ &+ a_n(x - x_0) \cdots (x - x_{n-1}) \end{aligned}$$

so that we can write

$$p_n(x) = a_0 + (x - x_0)(a_1 + (x - x_1)(a_2 + \cdots + (x - x_{n-1})a_n) \ldots))$$

Example 8.6 For example, let us create a quadratic polynomial that interpolates the sine function inside the interval $[0, \pi]$, with nodes that are evenly spaced. The function $f(x) = \sin x, x_i = 0, \frac{1}{2}\pi, \pi$ and $y_i = 0, 1, 0$. Subsequently

$$a_0 = 0 \quad \Rightarrow p_0(x) = 0$$

$$a_1 = \frac{\sin\frac{1}{2}\pi - p_0\left(\frac{1}{2}\pi\right)}{\left(\frac{1}{2}\pi - 0\right)} = \frac{2}{\pi} \Rightarrow p_1(x) = \frac{2}{\pi}x$$

$$a_2 = \frac{\sin\pi - p_1(\pi)}{(\pi - 0)\left(\pi - \frac{1}{2}\pi\right)} = -\frac{4}{\pi^2} \Rightarrow p_2(x) = \frac{2}{\pi}x - \frac{4}{\pi^2}x\left(x - \frac{1}{2}\pi\right)$$

It should be checked that each interpolating polynomial does interpolate at the appropriate points.

The coefficients a_k are called divided differences, and we can construct an algorithm for computing them, based on the formula (8.14). It is also possible to slightly modify this algorithm to allow the user to update an existing set of divided difference coefficients by adding new points to the set of nodes.

An alternative method for determining the divided difference coefficients, which has been historically employed, involves utilising a divided difference table. Given its simplicity in facilitating manual calculations, it is pertinent to consider this method in this context.

To start, we establish clear definitions for certain terms and symbols. The function values, which correspond to a set of nodes, will be denoted as the *zero*-th division differences. In this context, we shall represent them as $f_0(x_k)$. The first divisions are subsequently derived from the *zero*-th values in accordance with

$$f_1(x_k) = \frac{f_0(x_{k+1}) - f_0(x_k)}{x_{k+1} - x_k}$$

and then the second divided differences are given by

$$f_2(x_k) = \frac{f_1(x_{k+1}) - f_1(x_k)}{x_{k+2} - x_k}$$

and so on. The general formula is

$$f_j(x_k) = \frac{f_{j-1}(x_{k+1}) - f_{j-1}(x_k)}{x_{k+j} - x_k}$$

By employing the specified vocabulary and notation, it is possible to design a series of tables that may be easily used for the computation of interpolating polynomials.

Example 8.7 Take $f(x) = \log 2x$ as the function to be interpolated, using the nodes $x_0 = 1$, $x_1 = 2, x_2 = 4$. We begin by arranging the nodes and corresponding functional data in two columns, as in Table 8.2. Note that we write the functional data using the divided difference notation.

Table 8.2 Initial table for divided differences.

k	x_k	$f_0(x_k)$
0	1	0
1	2	1
2	4	2

The third column of the table is formed by the first divided differences. Thus,

$$1 = \frac{f(x_1) - f(x_0)}{x_1 - x_0} = \frac{1 - 0}{2 - 1} = f_1(x_0)$$

$$\frac{1}{2} = \frac{f(x_2) - f(x_1)}{x_2 - x_1} = \frac{2 - 1}{4 - 2} = f_1(x_1)$$

so we get Table 8.3.

The fourth (and, in this case, final) column is formed by the second divided difference:

$$-\frac{1}{6} = \frac{\frac{1}{2} - 1}{4 - 1} = f_2(x_0)$$

see Table 8.4.

Table 8.3 Table for divided differences after computation of first differences

k	x_k	$f_0(x_k)$	$f_1(x_k)$
0	1	0	
			1
1	2	1	
			1/2
2	4	2	

Table 8.4 Table for divided differences after computation of second differences.

k	x_k	$f_0(x_k)$	$f_1(x_k)$	$f_2(x_k)$
0	1	0		
			1	
1	2	1		$-\frac{1}{6}$
			$\frac{1}{2}$	
2	4	2		

The row of table values provides the divided difference coefficients, which are arranged in ascending order based on the nodes. Therefore

$$p_2(x) = 0 + (1)(x-1) - \frac{1}{6}(x-1)(x-2) = -\frac{1}{6}(x-1)(x-2)$$

The values generated by the pseudocode are similar. The coefficients obtained from the divided difference are presented in the bottom diagonal row, where the nodes are numbered in reverse order. Therefore, we also possess

$$p_2(x) = 2 + \left(\frac{1}{2}\right)(x-4) - \frac{1}{6}(x-4)(x-2) = -\frac{1}{6}(x-1)(x-8)$$

Adding a node to the data set and constructing the new polynomial is a straightforward process, as indicated by these tables. Let us incorporate the additional node $x_3 = \frac{1}{2}$ into our existing illustration. The initial layout is presented in Table 8.5. (Please be aware that the nodes are not needed to be organised in ascending order, or any specific order.

Table 8.5 Table for divided differences with new node added

k	x_k	$f_0(x_k)$	$f_1(x_k)$	$f_2(x_k)$
0	1	0		
			1	
1	2	1		$-\frac{1}{6}$
			$\frac{1}{2}$	
2	4	2		
3	$\frac{1}{2}$	-1		

Computing forward we get the (new) final result in Table 8.6, from which we conclude that the interpolating polynomial is

$$\begin{aligned} p_3(x) = 0 + (1)(x-1) - \frac{1}{6}(x-1)(x-2) \\ + \frac{1}{7}(x-1)(x-2)(x-4) \end{aligned}$$

Table 8.6 Table for divided differences after computation of third differences

k	x_k	$f_0(x_k)$	$f_1(x_k)$	$f_2(x_k)$	$f_3(x_k)$
0	1	0			
			1		
1	2	1		$-\frac{1}{6}$	
			$\frac{1}{2}$		$\frac{1}{7}$
2	4	2		$-\frac{5}{21}$	
			$\frac{6}{7}$		
3	$\frac{1}{2}$	-1			

Example 8.8 Using the following table find $f(x)$ as a polynomial in x.

x	$f(x)$
-1	3
0	-6
3	39
6	822
7	1611

Solution. The divided difference table is

Table 8.7 Table for divided differences for Example 8.8

x	$f(x)$				
-1	3				
		-9			
0	-6		6		
		15		5	
3	39		41		1
		261		13	
6	822		132		
		789			
7	1611				

Hence,

$$f(x) = 3 + (x+1)(-9) + x(x+1)(x-3)(5) + x(x+1)(x-3)(x-6) = x^4 - 3x^3 + 5x^2 - 6$$

8.5 Piecewise Linear Interpolation

Frequently, polynomial interpolation proves to be inadequate as a means of approximation. This assertion has validity when we persist in allowing the order of the polynomial to progressively increase. Nevertheless, by maintaining a consistent order of the polynomial and employing distinct polynomials over varying intervals, with the intervals gradually decreasing in length, interpolation can serve as a highly precise and potent approximation technique.

Let us commence by employing linear interpolation within the interval $I = [a, b]$, resulting in

$$p_1(x) = f(a) + f[a,b](x-a) = f(a) + \frac{f(b) - f(a)}{b-a}(x-a)$$
$$= \frac{x-b}{a-b}f(a) + \frac{x-b}{b-a}f(b) \quad (8.15)$$

As previously mentioned, a quadratic interpolation polynomial can be derived using the nodes $x_0 = a,\ x_2 = b$, and $x_0 < x_1 < x_2$. Conversely, the function $f(x)$ can be approximated using two linear polynomials: one in the range $[x_0, x_1]$ and the other in $[x_1, x_2]$. These polynomials are characterised as

$$p_{1,1}(x) = \frac{\mathrm{x} - x_1}{x_0 - x_1} f(x_0) + \frac{\mathrm{x} - x_1}{x_0 - x_1} f(x_1),$$

$$p_{1,2}(x) = \frac{\mathrm{x} - x_2}{x_1 - x_2} f(x_1) + \frac{\mathrm{x} - x_1}{x_2 - x_1} f(x_2),$$

and the interpolating polynomial is given by

$$P(x) = \begin{cases} p_{1,1}(x), & x \in [x_0, x_1] \\ p_{1,2}(x), & x \in [x_1, x_2] \end{cases} \tag{8.16}$$

P(x) is a continuous function within the interval $[a, b]$, interpolating $f(x)$, and is linear within the intervals $[a, x_1]$ and $[x_1, b]$.

The term for such a polynomial is piecewise linear polynomial. While piecewise linear interpolation maintains continuity, it lacks differentiability at the nodes and approximates f(x) inadequately. Our objective is to identify a smooth interpolation function that provides an improved approximation of f(x). This is something that can be accomplished via spline interpolation.

Example 8.9 Consider the following 11 data points:

$(-1.00, 0.038),\ (-0.80, 0.058),\ (-0.60, 0.100),\ (-0.4, 0.200),\ (-0.20, 0.500),$
$(0.00, 1.000), (0.20, 0.500), (0.40, 0.200), (0.60, 0.100), (0.80, 0.0580), (1.00, 0.038)$

which were generated using the Runge function. Find the explicit representation of the piecewise linear interpolating function for these data points.

Solution. There are 11 data points surrounding 10 intervals. For each interval i, a set of coefficients a_i and b_i are required for the linear interpolation. For the first interval, the coefficients are:

$$a_1 = \frac{y_1 - y_2}{x_1 - x_2} = \frac{0.038 - 0.058}{-1 - (-0.8)} = 0.1$$
$$b_1 = \frac{y_2 x_1 - y_1 x_2}{x_1 - x_2} = \frac{0.058(-1) - 38(-0.8)}{-1 - (-0.8)} = 0.138$$

Similarly,

$$a_2 = \frac{y_2 - y_3}{x_2 - x_3} = \frac{-0.1}{-0.8 - (-0.6)} = 0.21$$
$$b_2 = \frac{y_3 x_2 - y_2 x_3}{x_2 - x_3} = \frac{0.1(-0.8) - 0.058(-0.6)}{-0.8 - (-0.6)} = 0.226$$

Repeating for the other coefficients, the required explicit equation has the form:

$$y = \begin{cases} 0.1x + 0.138, & x \in [-1,] \\ 0.21x + 0.226, & x \in [-0.8, -0.6] \\ 0.5x + 0.4, & x \in [-0.6, -0.4] \\ 1.5x + 0.8, & x \in [-0.4, -0.2] \\ 2.5x + 1, & x \in [-0.2, 0] \\ -2.5x + 1, & x \in [0, 0.2] \\ -1.5x + 0.8, & x \in [0.2, 0.4] \\ -0.5x + 0.4, & x \in [0.4, 0.6] \\ -0.21x + 0.226, & x \in [0.6, 0.8] \\ -0.1x + 0.138, & x \in [0.8, 1] \end{cases}$$

Notice that for the procedure to work, the x components of the data points have to satisfy: $x_1 < x_2 < \cdots x_k$. The following Mathematica code utilizes the required piecewise linear function.

8.6 Derivation of Cubic Splines

The primary aim of cubic splines is to obtain a third-order polynomial for every interval between knots, as exemplified by the equation

$$f_i(x) = a_i x^3 + b_i x^2 + c_i x + d_i \tag{8.17}$$

Hence, in the case of $n + 1$ data points ($i = 0,1,2,\ldots,n$), there are n gaps and, as a result, $4n$ undetermined constants that need to be assessed. Similar to quadratic splines, the evaluation of the unknowns necessitates the use of 4n conditions. The following items are:

1. According to the $2n-2$ requirements, it is necessary for the function values to be equal at the inner knots.
2. Two requirements must be met for the first and last functions to pass through the end points.
3. According to the $n-1$ requirements, it is necessary for the first derivatives at the inner knots to be equal.
4. According to the above constraints, it is necessary for the second derivatives at the inner knots to be equal $(n-1)$.
5. In the end knots, the second derivatives exhibit a value of zero, satisfying two requirements.

In this section, we will look at the methodology for obtaining the interpolation of a function $f(x)$ using spline functions as opposed to polynomials. In order to maintain simplicity, we limit our analysis solely to cubic splines. The spline interpolation $s(x)$ of a function $f(x)$ is constructed as follows:

Step 1: Let $M_1, \dots, M_n$, and M_i be denoted as $s(x_i)$, where $i = 0,1,\dots,n$. The first step involves obtaining $s(x)$ in terms of the unknowns M_i.

Step 2: As $s(x)$ is cubic on each $[x_{i-1}, x_i]$, the function $s''(x)$ is linear on the interval such that

$$s''(x_{i-1}) = M_{i-1}, \qquad s''(x_i) = M_i$$

Therefore, it is given by

$$s''(x) = \frac{(x_i - x)M_{i-1} + (x - x_{i-1})M_i}{x_i - x_{i-1}}, \quad x_{i-1} \le x \le x_i \tag{8.18}$$

Integrating (8.18) two times with respect to x, we get

$$s(x) = \frac{(x_i - x)^3 M_{i-1}}{6(x_i - x_{i-1})} + \frac{(x - x_{i-1})^3 M_i}{6(x_i - x_{i-1})} + K_1 x + K_2$$

The integration constants and K_2 are found by applying the conditions $s(x_{i-1}) = f(x_{i-1})$ and $s(x_i) = f(x_i)$. We offer

$$K_1 = \frac{f(x_i) - f(x_{i-1})}{x_i - x_{i-1}} - \frac{(M_i - M_{i-1})(x_i - x_{i-1})}{6}$$

$$K_2 = \frac{x_i f(x_{i-1}) - x_{i-1} f(x_i)}{x_i - x_{i-1}} - \frac{(M_{i-1}x_i - M_i x_{i-1})(x_i - x_{i-1})}{6}$$

Substituting these values in the above equation, we get

$$\begin{aligned} s(x) = {} & \frac{(x_i - x)^3 M_{i-1} + (x - x_{i-1})^3 M_i}{6(x_i - x_{i-1})} \\ & + \frac{(x_i - x) f(x_{i-1}) + (x_i - x_{i-1}) f(x_i)}{(x_i - x_{i-1})} \\ & - \frac{1}{6}((x_i - x_{i-1})[((x_i - x)M_{i-1} \\ & + (x - x_{i-1})M_i], \quad x_{i-1} \le x \le x_i \end{aligned} \tag{8.19}$$

The equation (8.19) is applicable to all intervals $[x_1, x_2], \ldots, [x_{n-1}, x_n]$. Due to the interpolating requirement $s(x_i) = f(x_i)$, the formulas for adjacent intervals $[x_{i-1}, x_i]$ and $[x_i, x_{i+1}]$ will converge at their shared point $x = x_i$. This statement suggests that the function $s(x)$ will exhibit continuity across the full interval $[a, b]$. Formula (8.18) for $s''(x)$ indicates that it is continuous on the interval $[a, b]$.

Step 3: The remaining task is to determine the values of M_i for all i values, ranging from 0 to n. The continuity of $s'(x)$ over $[a, b]$ is established by ensuring that the formula for $s'(x)$ on $[x_{i-1}, x_i]$ and $[x_i, x_{i+1}]$ yields the same value at their shared point $x = x_i$, for $i = 1, 2, \ldots, n-1$. After simplification, we get the system of linear equations for $i = 1, 2, \ldots, n-1$

$$\begin{aligned} & \frac{x_i - x_{i-1}}{6} M_{i-1} + \frac{x_{i+1} - x_{i-1}}{3} M_i + \frac{x_{i+1} - x_i}{6} M_{i+1} \\ & = \frac{f(x_{i+1}) - f(x_i)}{x_{i+1} - x_i} - \frac{f(x_i) - f(x_{i-1})}{x_i - x_{i-1}} \end{aligned} \tag{8.20}$$

These $n - 1$ equations together with the assumption that

$$M_0 = M_n = 0 \tag{8.21}$$

leads to the values of $M_0, M_1, \ldots, M_n$ and hence to the interpolation function $s(x)$. A spline constructed above is called a natural spline.

Example 8.10 Calculate the natural cubic spline interpolating the data

$$\left\{(1,1), \left(2, \frac{1}{2}\right), \left(3, \frac{1}{3}\right), \left(4, \frac{1}{4}\right)\right\}$$

Solution. The number of points is $n = 4$ and all $x_i - x_{i-1} = 1$. The system (8.20) together with $M_0 = M_3 = 0$ becomes

$$\frac{2}{3}M_2 + \frac{1}{6}M_3 = \frac{1}{3}, \qquad \frac{1}{6}M_1 + \frac{2}{3}M_3 = \frac{1}{12}$$

which gives $M_2 = \frac{1}{2}$, $M_3 = 0$. Substituting these values into (8.19), we get

$$s(x) = \begin{cases} \frac{1}{12}x^3 - \frac{1}{4}x^2 - \frac{1}{3}x + \frac{3}{2}, & 1 \le x \le 2 \\ -\frac{1}{12}x^3 + \frac{3}{4}x^2 - \frac{7}{3}x + \frac{17}{6}, & 2 \le x \le 3 \\ -\frac{1}{3}x + \frac{3}{2}, & 3 \le x \le 4 \end{cases}$$

Example 8.11 Fit cubic splines to the following data (Table 8.8). Utilize the results to estimate the value at $x = 5$.

Table 8.8 Data to be fit with spline functions

x	$f(x)$
3.0	2.5
4.5	1.0
7.0	2.5
9.0	0.5

Solution. For the first interior knot, the following data are used:

$$\begin{array}{ll} x_0 = 3 & f(x_0) = 2.5 \\ x_1 = 4.5 & f(x_1) = 1 \\ x_2 = 7 & f(x_2) = 2.5 \end{array}$$

These values can be substituted to yield

$$(4.5 - 3)f''(3) + 2(7 - 3)f''(4.5) + (7 - 4.5)f''(7)$$

$$= \frac{6}{7 - 4.5}(2.5 - 1) + \frac{6}{4.5 - 3}(2.5 - 1)$$

Because of the natural spline condition, $f''(3) = 0$, and the equation reduces to

$$8f''(4.5) + 2.5f''(7) = 9.6$$

In a similar fashion,

$$2.5f''(4.5) + 9f''(7) = -9.6$$

These two equations can be solved simultaneously for

$$f''(4.5) = 1.67909$$
$$f''(7) = -1.53308$$

These values can then be substituted to yield

$$f_1(x) = \frac{1.67909}{6(4.5 - 3)}(x - 3)^3 + \frac{2.5}{4.5 - 3}(4.5 - x) + \left[\frac{1}{4.5 - 3} - \frac{1.67909(4.5\ 2\ 3)}{6}\right](x - 3)$$

Or

$$f_1(x) = 0.186566(x - 3)^3 + 1.666667(4.5 - x) + 0.246894(x - 3)$$

This equation is the cubic spline for the first interval. Similar substitutions can be made to develop the equations for the second and third intervals:

$$f_2(x) = 0.111939(7 - x)^3 - 0.102205(x - 4.5)^3 - 0.299621(7 - x) + 1.638783(x - 4.5)$$

And

$$f_3(x) = -0.127757(9 - x)^3 + 1.761027(7 - x) + 0.25(x - 7)$$

The three equations can then be employed to compute values within each interval. For example, the value at $x = 5$, which falls within the second interval, is calculated as

$$f_2(5) = 0.111939(7 - 5)^3 - 0.102205(5 - 4.5)^3 - 0.299621(7 - 5) + 1.638783(5 - 4.5) = 1.102886$$

Other values are computed, and the results are plotted in Figure 8.4.

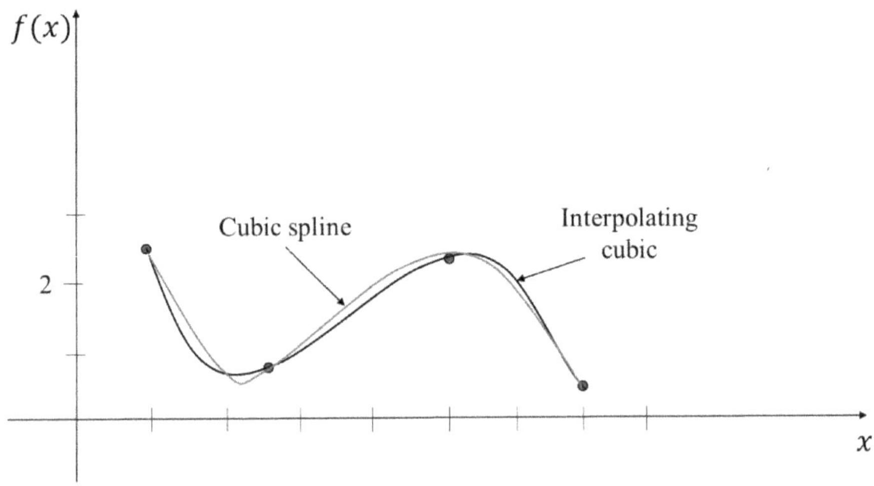

Figure 8.4 cubic spline fits of a set of four points with a cubic interpolating polynomial.

8.7 Inverse Interpolation

In the majority of interpolation settings, the $f(x)$ and x values are commonly referred to as the dependent and independent variables, respectively. Consequently, the values of the x's are usually evenly distributed. An illustrative instance is a tabular representation of values obtained for the function $f(x) = \frac{1}{x}$,

x	1	2	3	4	5	6	7
$f(x)$	1	0.5	0.3333	0.25	0.2	0.1667	0.1429

Consider an example where it is necessary to use identical data, although you are provided with a value for $f(x)$ and are tasked with determining the matching value of x. For the above dataset, let us consider the scenario where you were tasked with identifying the value of x that corresponds to the function $f(x) = 0.3$.

In this particular scenario, due to the accessibility and ease of manipulation of the function, the accurate solution may be directly ascertained as $x = \frac{1}{0.3} = 3.3333$.

The issue at hand is commonly referred to as inverse interpolation. In the context of a more intricate scenario, one may be inclined to manipulate the values of $f(x)$ and x by plotting x against $f(x)$ and use a method such as Lagrange interpolation to ascertain the outcome. Regrettably, the reversal of variables does not provide a guarantee of equal spacing among the values along the new abscissa, denoted as $f(x)$'s. Indeed, in numerous instances, the values will be "limited."

In other words, the data will have a logarithmic scale, with certain nearby points closely clustered together while others are far apart. For instance, for $f(x) = \frac{1}{x}$ the result is

$f(x)$	0.1429	0.1667	0.2	0.25	0.3333	0.5	1
x	7	6	5	4	3	2	1

The uneven spacing on the x-axis can frequently cause oscillations in the resulting interpolating polynomial. This phenomenon can also manifest in polynomials of lower orders.

Another approach is to fit a nth-order interpolating polynomial, $f_n(x)$, to the original data (i.e., with $f(x)$ versus x). Typically, due to the uniform spacing of the x's, this polynomial is unlikely to exhibit ill-conditioning. To solve your problem, we need to determine the value of x that will make this polynomial equal to the given $f(x)$. Therefore, the problem of interpolation can be simplified to a problem of finding roots.

As an example, one possible solution for the previously mentioned issue is to fit a quadratic polynomial to the set of three points: $(2, 0.5)$, $(3, 0.3333)$, and $(4, 0.25)$. The outcome would be

$$f_2(x) = 1.08333 - 0.375x + 0.041667x^2$$

The answer to the inverse interpolation problem of finding the x corresponding to $f(x) = 0.3$ would therefore involve determining the root of

$$0.3 = 1.08333 - 0.375x + 0.041667x^2$$

For this simple case, the quadratic formula can be used to calculate

$$x = \frac{0.375 \pm \sqrt{(-0.375)^2 - 4(0.041667)0.78333}}{2(0.041667)} = \begin{matrix} 5.704158 \\ 3.295842 \end{matrix}$$

Hence, the second root, 3.296, serves as a reliable approximation for the actual value of 3.333.

8.8 Multidimensional interpolation

Consider a scenario where we are estimating the value of a function that involves two variables, denoted as $z = f(x, y)$.

The known values are represented by the equation $z_{ij} = f(x_i, y_j)$, where i ranges from 0 to n and j ranges from 0 to n. It should be noted that the places (x, y) where the function $f(x, y)$ is known are located on a grid within the $x - y$ plane.
Consider the interpolating function $z = g(x, y)$, which satisfies $z_{ij} = f(x_i, y_j)$.

To determine the value of g at the point (x, y), a two-dimensional interpolation can be employed. This involves initially conducting $n + 1$ one-dimensional interpolations in y to ascertain the value of g at the $n + 1$ points $(x_0, y), (x_1, y), \ldots, (x_n, y)$. Subsequently, a single one-dimensional interpolation in x is performed to determine the value of g at (x, y).

To clarify, the process of two-dimensional interpolation on a grid of dimensions $(n + 1)(n + 1)$ involves initially conducting $n + 1$ one-dimensional interpolations on the value y, followed by a solitary one-dimensional interpolation on the value x. The process of interpolating in two dimensions can be extended to higher dimensions.

8.9 Exercises

1. Obtain Lagrange interpolation formula for equally spaced nodes.

2. Using Lagrange interpolation formula, express the rational function $f(x) = \sin \pi x$, choosing the points $x_0 = 0, x_1 = \frac{1}{6}, x_3 = \frac{1}{2}$

3. Find a cubic polynomial using Lagrange's formula for the data:

x	-2	-1	1	3
$f(x)$	-1	3	-1	19

4. Use Lagrange interpolation formula to find a quadratic polynomial $P_2(x)$ that interpolates the function $f(x) = e^{-x^2}$ at $x_0 = -1, x_1 = 0$ and $x_2 = 1$. Further, find the value of $P_2(-0.9)$ with rounding to six decimal places after decimal point and compare the value with the true value $f(-0.9)$ of same figure.

5. Given a table of values of the function $f(x)$

x	321.0	322.8	324.2	325.0
$f(x)$	2.50651	2.50893	2.51081	2.51188

6. The functions

$$l_k = \prod_{i=0, i\neq k}^{n} \frac{x - x_i}{x_k - x_i}, k = 0, \dots, n$$

are the weight polynomials of the corresponding nodes and are often called Lagrange multipliers. Prove that for any $n \geq 1, \sum_{k=0}^{n} l_k(t) = 1$.

7. Let $x_k \in [a, b], k = 0,1, \dots, n$ be $n + 1$ distinct nodes and let $f(x)$ be a continuous function on $[a, b]$. Show that for $x \neq x_k, k = 0,1, \dots, n$, the Lagrange interpolating polynomial can be represented in the form

$$P_n(x) = \omega(x) \sum_{k=0}^{n} \frac{f(x_k)}{(x - x_k)\omega'(x_k)}$$

8. For the given functions $f(x)$, let $x_0 = 0, x_1 = 0.6$, and $x_2 = 0.9$. Construct interpolation polynomials of degree at most one and at most two to approximate $f(0.45)$, and find the absolute error.

a) $f(x) = \cos x$
b) $f(x) = \sqrt{1 + x}$
c) $f(x) = \ln(x + 1)$
d) $f(x) = \tan x$

9. Use Newton the forward-difference formula to construct interpolating polynomials of degree one, two, and three for the following data. Approximate the specified value using each of the polynomials.

a) $f(-\frac{1}{3})$ if $f(-0.75) = -0.07181250, f(-0.5) = -0.02475000, f(-0.25) = 0.33493750, f(0) = 1.10100000$

b) $f(0.25)$ if $f(0.1) = -0.62049958$, $f(0.2) = -0.28398668$, $f(0.3) = 0.00660095, f(0.4) = 0.24842440$

10. Use the Newton forward-difference formula to construct interpolating polynomials of degree one, two, and three for the following data. Approximate the specified value using each of the polynomials.

a) $f(0.43)$ if $f(0) = 1, f(0.25) = 1.64872, f(0.5) = 2.71828$, $f(0.75) = 4.48169$

b) $f(0.18)$ if $f(0.1) = -0.29004986$, $f(0.2) = -0.56079734$, $f(0.3) = -0.81401972, f(0.4) = -1.0526302$

11. Use the Newton backward-difference formula to construct interpolating polynomials of degree one, two, and three for the following data. Approximate the specified value using each of the polynomials.

a) $f(-\frac{1}{3})$ if $f(-0.75) = -0.07181250, f(-0.5) = -0.02475000, f(-0.25) = 0.33493750, f(0) = 1.10100000$

b) $f(0.25)$ if $f(0.1) = -0.62049958$, $f(0.2) = -0.28398668$, $f(0.3) = 0.00660095, f(0.4) = 0.24842440$

12. a) Approximate $f(0.05)$ using the following data and the Newton forward-difference formula:

x	0	0.2	0.4	0.6	0.8
$f(x)$	1	1.22140	1.49182	1.82212	2.22554

b) Use the Newton backward-difference formula to approximate $f(0.65)$.

13. For the function data given in the table below, fit a polynomial using Newton interpolation formula and find the value of $f(2.5)$.

x	−3	−1	0	3	5
$f(x)$	−30	−22	−12	330	3458

14. Calculate the nth divided difference of $f(x) = \frac{1}{x}$.

15. Construct a natural cubic spline to approximate $f(x) = e^{-x}$ by using the values given by $f(x)$ at $x = 0, 0.25, 0.75,$ and 1.0. Integrate the spline over $[0, 1]$, and compare the result to $\int_0^1 e^{-x}\,dx = 1 - \frac{1}{e}$. Use the derivatives of the spline to approximate $f'(0.5)$ and $f''(0.5)$. Compare the approximations to the actual values.

16. Obtain the cubic spline approximation for the function given in the tabular form

x	0	1	2	3
$f(x)$	1	2	33	244

17. Let $p_3(x)$ denote the cubic polynomial interpolating $f(x)$ at the evenly spaced points $x_j = x_0 + jh,\ j = 0, 1, 2, 3$. Assuming $f(x)$ is sufficiently differentiable, bound the error in using $p_3{}'(x)$ as an approximation to $f'(x), x_0 \le x \le x_3$.

18. Consider finding a cubic spline interpolating function for the data

x	0	1	2	2.5	3	4
y	1.4	0.6	1	0.65	0.6	1

Graph the resulting function $s(x)$. Compare it to the use of piecewise linear interpolation, connecting the successive points (x_i, y_i) by line segments.

19. Find a cubic polynomial which fits the data: $(-2, -12), (-1, -8), (2,3)$ and $(3,5)$.

20. Establish Newton's divided difference interpolation formula and give an estimate of the reminder term. Deduce Newton's forward and backward-difference formulae as particular cases.

21. Use the cubic spline interpolation scheme to find the interpolating function given the following 5 data points:

$(-1, 0.038)\ (-0.8, 0.058), (-0.60, 0.10), (-0.4, 0.20), (-0.2, 0.5)$

CHAPTER 9

Numerical Integration

9.1 Introduction

This chapter provides an analysis and derivation of numerical techniques for definite integral evaluation. Integration is, in general, the process of adding up segments or pieces to determine the total. When dx represents a small displacement or change along the direction x, the integration process yields a function $g(x)$ whose derivative is equal to the function $f(x)$. Therefore, the integral of $f(x)dx$ represents the product of $f(x)$ and dx. Numerical integration involves the use of computational methods to approximate and evaluate definite integrals. Numerical integration is a collection of algorithms used to calculate the value of a definite integral. It is also sometimes used to refer to the numerical solution of differential equations.

Numerical quadrature, also known as quadrature, is essentially synonymous with numerical integration, particularly in the context of one-dimensional integrals. There is a difference in terminology among authors when it comes to numerical integration over multiple dimensions. Some use the term "cubature" while others consider "quadrature" to encompass higher-dimensional integration. Numerical integration methods involve combining evaluations of the integrand to obtain an approximation of the integral. The function is calculated at a finite set of points known as integration points. A weighted sum of these values is then utilised to approximate the integral. The integration points and weights vary depending on the chosen method and the desired level of accuracy for the approximation.

When the integrand $f(x)$ is not integrable or when the function is only known by its values at a finite number of points, it becomes impossible to evaluate a definite integral. Nevertheless, the integral's value can be determined through numerical methods. There are two distinct numerical methods available for the evaluation of a definite integral, which are based on the following formula

$$\int_a^b f(x)dx \quad (9.1)$$

These methods are known as Newton-Cotes quadrature and Gaussian quadrature.

9.2 A Review of Definite Integrals

In order to calculate the definite integral of a function, we begin by establishing the mesh points (or grid points) $x_i^{(n)}$ based on the equation

$$a = x_0^{(n)} < x_1^{(n)} < x_2^{(n)} \dots < x_{n-1}^{(n)} < x_n^{(n)} = b \quad (9.2)$$

Next, we select the evaluation points n_i from within the corresponding subinterval, specifically $n_i \in [x_{i-1}, x_i]$, where $1 < i < n$, without any specific criteria. Now, let's assume that

$$\lim_{n\to\infty} \left[\max_{1\le i\le n} \left(x_i^{(n)} - x_{i-1}^{(n)} \right) \right] = 0$$

It can be demonstrated, given certain conditions on f, that the limit is reached when the largest distance between adjacent points approaches zero with an increasing number of points, then

$$L = \lim_{n\to\infty} \sum_{i=1}^{n} f(n_i) \left(x_i^{(n)} - x_{i-1}^{(n)} \right) \quad (9.3)$$

Importantly, the value of this limit is not affected by the specific choices made for the mesh points $\left\{ x_i^{(n)} \right\}$ and evaluation points $\{n_i\}$. When this occurs, the value that represents this limit is referred to as the definite integral of f , and we write,

$$L = I(f) = \int_a^b f(x)dx$$

It should be noted that this integral construction is really simple. The essential requirements are that the function is continuous, the sequence of meshes is sufficiently fine, and the evaluation points are chosen freely within each subinterval. A Riemann sum, named after the German mathematician Georg Bernhard Riemann, is a concise representation denoted as $R_n(f)$.

$$R_n(f) = \sum_{i=1}^{n} f(n_i)\left(x_i^{(n)} - x_{i-1}^{(n)}\right) \tag{9.4}$$

The trapezoidal rule, also referred to as the trapezoid rule or trapezium rule, is a numerical analysis method used to estimate the value of a definite integral. The trapezoidal rule can be seen as the outcome achieved by averaging the left and right Riemann sums and is occasionally defined in this manner.

Example 9.1 Show that the function $y(t) = 1 - t$ is a solution of the integral equation $t = \int_0^t (e^{(t-u)}(1-u)du$

Solution. We have

$$\int_0^t (e^{(t-u)}y(u)du = \int_0^t (e^{(t-u)}(1-u)du$$

$$= e^t \int_0^t (e^{(t-u)}(1-u)du = e^t\left[-e^{-u}(1-u) + \int e^{-u}(-1)\,du\right]$$

$$= t, \text{ on simplification.}$$

9.3 Newton-Cotes Integration Formulas

The Newton-Cotes formulas are widely used in numerical integration schemes. These methods rely on the approach of substituting a complex function or a table of data with a simpler function that can be easily integrated:

$$I = \int_a^b f(x)dx \cong \int_a^b f_n(x)dx \tag{9.5}$$

Where $f_n(x) = a_0 + a_1x + \cdots + a_{n-1}x^{n-1} + a_nx^n$

where n is the order of the polynomial. An alternative method for approximating the integral involves using a series of polynomials that are applied to the function or data in segments of constant length.

Higher-order polynomials can also be used for the same purpose. There are both closed and open forms of the Newton-Cotes formulas that can be utilised. The closed forms refer to situations where the data points at the boundaries of the integration limits are known. The open forms of integration have limits that extend beyond the range of the available data. Definite integration typically does not involve the use of Open Newton-Cotes formulas.

9.4 Trapezoidal Rule

The trapezoidal rule is one of the first Newton-Cotes closed integration formulas. This corresponds to the case where the polynomial in Equation (9.5) is of first order.

$$I = \int_a^b f(x)dx \cong \int_a^b f_1(x)dx$$

We know, a straight line can be represented as

$$f_1(x) = f(a) + \frac{f(b) - f(a)}{b - a}(x - a) \tag{9.6}$$

The integral of $f(x)$ between the limits a and b can be estimated by finding the area under this straight line:

$$I = \int_a^b \left[f(a) + \frac{f(b) - f(a)}{b - a}(x - a)\right] dx$$

The result of the integration is

$$I = (b - a)\frac{f(a) + f(b)}{2} \tag{9.7}$$

which is called the trapezoidal rule.

9.4.1 Derivation of Trapezoidal Rule

Before integration, Eq. (9.6) can be expressed as

$$f_1(x) = \frac{f(b) - f(a)}{b - a}x + f(a) - \frac{af(b) - af(a)}{b - a}$$

Grouping the last two terms gives

$$f_1(x) = \frac{f(b) - f(a)}{b - a}x + f(a) - \frac{bf(a) - af(b) - af(b) + af(a)}{b - a}$$

Or

$$f_1(x) = \frac{f(b) - f(a)}{b - a}x + \frac{bf(a) - af(b)}{b - a}$$

which can be integrated between $x = a$ and $x = b$ to yield

$$I = \frac{f(b) - f(a)}{b - a}\frac{x^2}{2} + \frac{bf(a) - af(b)}{b - a}x \Big|_a^b$$

This result can be evaluated to provide

$$I = \frac{f(b) - f(a)}{b - a}\frac{(b^2 - a^2)}{2} + \frac{bf(a) - af(b)}{b - a}(b - a)$$

Now, since $b^2 - a^2 = (b - a)(b + a)$,

$$I = [f(b) - f(a)]\frac{b + a}{2} + bf(a) - af(b)$$

The result is obtained by multiplying and combining terms

$$I = (b - a)\frac{f(b) + f(a)}{2}$$

which is the formula for the trapezoidal rule.

From a geometric perspective, the trapezoidal rule can be seen as an approximation of the area of a trapezoid. This trapezoid is formed by the straight line connecting $f(a)$ and $f(b)$ in Figure 9-1. In geometry, the area of a trapezium can be calculated by multiplying the height by the average of the bases. In our case, the trapezium is oriented horizontally. Thus, the integral estimate can be expressed as

$$I \cong \text{width} \times \text{average height} \tag{9.8}$$

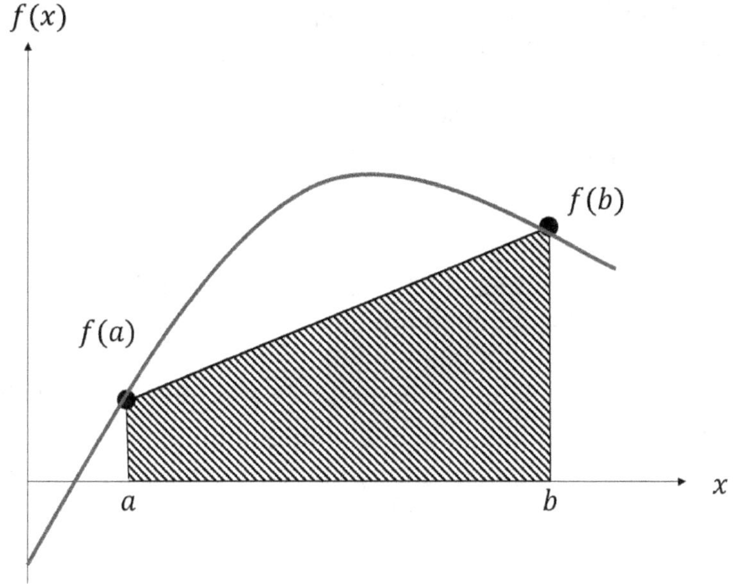

Figure 9.1 Graphical depiction of the trapezoidal rule

Or,

$$I \cong (b-a) \times \text{average height} \tag{9.9}$$

where, for the trapezoidal rule, the average height is the average of the function values at the end points, or $[f(a) + f(b)]/2$.

The general format of Eq. (9.9) includes all the Newton-Cotes closed formulas. Indeed, their only difference lies in the manner in which the average height is expressed.

When applying the integral under a straight-line segment to estimate the integral under a curve, it is evident that there is a possibility of a significant error. The local truncation error of a single application of the trapezoidal rule can be estimated as follows:

$$E_t = -\frac{1}{12} f''(\xi)(b-a)^3 \tag{9.10}$$

Where ξ is located within the interval from a to b. Equation (9.10) indicates that when the function being integrated is linear, the trapezoidal rule will yield precise results. However, when dealing with functions that have second- and higher-order derivatives (i.e., curvature), there is a possibility of encountering errors.

By integrating the forward Newton-Gregory interpolating polynomial, we get,

$$I = h\frac{f(a) + f(b)}{2} - \frac{1}{12}f''(\xi)h^3$$

Here the first term is the trapezoidal rule and the second is an approximation for the error.

Since $x_{j-1} \leq n_j \leq x_j$, $j = 1,\ldots,n$, the last sum is a Riemann sum; thus

$$E_n(f) = -\frac{h^2}{12}[f'(b) - f'(a)] \equiv \overline{E_n}(f) \tag{9.11}$$

The term $\overline{E_n}(f)$ is called an asymptotic error estimate for $E_n(f)$.

Example 9.2 Use Eq. (9.7) to numerically integrate

$$f(x) = 0.2 + 25x - 200x^2 + 675x^3 - 900x^4 + 400x^5$$

from $a = 0$ to $b = 0.8$. The exact value of the integral can be determined analytically to be 1.640533.

Solution. The function values

$$f(0) = 0.2$$
$$f(0.8) = 0.232$$

can be substituted into Eq. (9.7) to yield

$$I \cong 0.8\frac{0.2 + 0.232}{2} = 0.1728$$

which represents an error of

$$E_t = 1.640533 - 0.1728 = 1.467733$$

which corresponds to a percent relative error of $\varepsilon_t = 89.5\%$.
In real-life scenarios, we would lack any prior knowledge of the actual value. Thus, it is necessary to obtain an estimate of the error. In order to calculate this estimate, the second derivative of the function

over the given interval can be found by taking the derivative of the original function twice, resulting in

$$f''(x) = -400 + 4050x - 10{,}800x^2 + 8000x^3$$

The average value of the second derivative can be computed

$$f''(x) = \frac{\int_0^{0.8}(-400 + 4050x - 10{,}800x^2 + 8000x^3)dx}{0.8 - 0} = -60$$

which can be substituted into Eq. (9.10) to yield

$$E_a = -\frac{1}{12}(60)(0.8)^3 = 2.56$$

The error is similar in magnitude and direction to the true error. There is a discrepancy that arises due to the fact that, for an interval of this magnitude, the average second derivative may not provide an accurate approximation of $f''(\xi)$. Therefore, it is indicated that the error is estimated with the notation E_a, as opposed to being precisely determined with E_t.

Example 9.3 Use the trapezoidal rule to calculate

$$I = \int_0^{\pi} e^x \cos(x)dx$$

Solution. Here, $I = -\frac{e^{\pi}+1}{2} = -12.0703463164$

The values of I_n are given in Table 9.1, along with the true errors E_n and the asymptotic estimates E_n. Note that the errors decrease by a factor of 4 when n is doubled (and hence h is halved). This result was predictable from the multiplying factor of h^2 present in (9.10) and (9.11); when h is halved, h^2 decreases by a factor of 4. This example also shows that the trapezoidal rule is relatively inefficient when compared with other methods to be developed in this chapter.

Table 9.1 Trapezoidal rule for evaluating

n	I_n	E_n	Ratio	$\overline{E_n}$
2	−17.389259	5.32		4.96
4	−13.336023	1.27	4.20	1.24
8	−12.382162	$3.12E-1$	4.06	$3.10E-1$
16	−12.148004	$7.77E-2$	4.02	$7.76E-2$
32	−12.089742	$1.94E-2$	4.00	$1.94E-2$

64	−12.075194	4.85E − 3	4.00	4.85E − 3
128	−12.071558	1.21E − 3	4.00	1.21E − 3
256	−12.070649	3.03E − 4	4.00	3.03E − 4
512	−12.070422	7.57E − 5		7.57E − 5

In comparison to other techniques to be discussed in this chapter, this example also demonstrates the relative inefficiency of the trapezoidal rule.

Using the error estimate $\overline{E_n}(f)$, we can define an improved numerical integration rule:

$$CT_n(f) \equiv I_n(f) + \overline{E_n}(f)$$
$$= h\left[\frac{1}{2}f_0 + f_1 + \cdots + f_{n-1} + \frac{1}{2}f_n\right] - \frac{h^2}{12}[f'(b) - f'(a)] \tag{9.12}$$

$\overline{E_n}(f)$ is known for its high accuracy, which in turn makes $CT_n(f)$ significantly more precise compared to the trapezoidal rule.

Example 9.4 Use the trapezoidal rule to estimate $\int_0^1 x^2dx$ using four subintervals.

Solution. The endpoints of the subintervals consist of elements of the set $P = \left\{0, \frac{1}{4}, \frac{1}{2}, \frac{3}{4}, 1\right\}$ and $\Delta x = \frac{1-0}{4} = \frac{1}{4}$. Thus,

$$\int_0^1 x^2dx \approx \frac{1}{2}\cdot\frac{1}{4}\left(f(0) + 2f\left(\frac{1}{4}\right) + 2f\left(\frac{1}{2}\right) + 2f\left(\frac{3}{4}\right) + f(1)\right)$$
$$= \frac{1}{8}\left(0 + 2\cdot\frac{1}{16} + 2\cdot\frac{1}{4} + 2\cdot\frac{9}{16} + 1\right)$$
$$= \frac{11}{32}$$

Example 9.5 Approximate the integral

$$I = \int_0^1 \frac{dx}{1+x}$$

Solution. The true value is $I = \log(2) \approx 0.693147$. Using the trapezoidal rule (9.7), we get

$$I_T = \frac{1}{2}\left[1 + \frac{1}{2}\right] = \frac{3}{4} = 0.75$$

Therefore, the error is $I - I_T \approx -0.0569$.

9.5 Simpson's Rules

One other method to get a more precise approximation of an integral is to connect the points with higher-order polynomials, compared to using the trapezoidal method with finer segmentation. For example, if there exists an additional point located between $f(a)$ and $f(b)$, it is possible to connect the three points using a parabola (see Figure 9-2). When there are two equidistant points between $f(a)$ and $f(b)$, it is possible to connect all four points using a third-order polynomial (Fig. 9-3). The integrals under these polynomials give rise to formulas known as Simpson's rules. Simpson's rule is a method commonly employed in mathematics to estimate the precise value of a definite integral. In order to approximate the function over its interval (a, b), we divide it into n equal parts and use fitting polynomial identities obtained through Lagrange interpolation. By integrating these polynomials, we can obtain an approximation for the area under the curve of the function.

The fitting polynomial in Simpson's rule is a parabolic arc, rather than a straight line. In this calculation, the total area under the curve is divided into n equal subintervals to determine the area under the curve. Next, we employ three consecutive points within the interval and approximate them using parabolic curves. The derivation of Simpson's rule involves integrating a third-order Lagrange interpolating polynomial that is fitted to the function at three points that are equally spaced.

9.6 Simpson's 1/3 Rule

Simpson's 1/3 rule results when a second-order interpolating polynomial is substituted into Eq. (9.5):

$$I = \int_a^b f(x)dx \cong \int_a^b f_2(x)dx$$

If a and b are designated as x_0 and x_2 and $f_2(x)$ is represented by a second-order Lagrange polynomial, the integral becomes

$$I = \int_{x_0}^{x_2} \left[\frac{(x - x_1)(x - x_2)}{(x_0 - x_1)(x_0 - x_2)} f(x_0) + \frac{(x - x_0)(x - x_2)}{(x_1 - x_0)(x_1 - x_2)} f(x_1) + \frac{(x - x_0)(x - x_1)}{(x_2 - x_0)(x_2 - x_1)} f(x_2) \right] dx$$

After integration and algebraic manipulation, the following formula results:

$$I \cong \frac{h}{3} [f(x_0) + 4f(x_1) + f(x_2)] \tag{9.13}$$

where, $h = (b - a)/2$, for this case. This equation is known as Simpson's 1/3 rule.

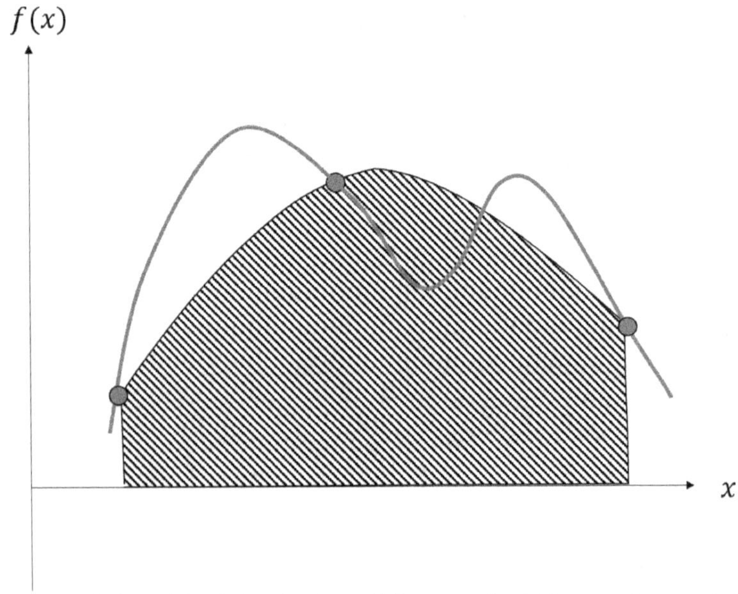

Figure 9.2 Graphical depiction of Simpson's 1/3 rule: It consists of taking the area under a parabola connecting three points.

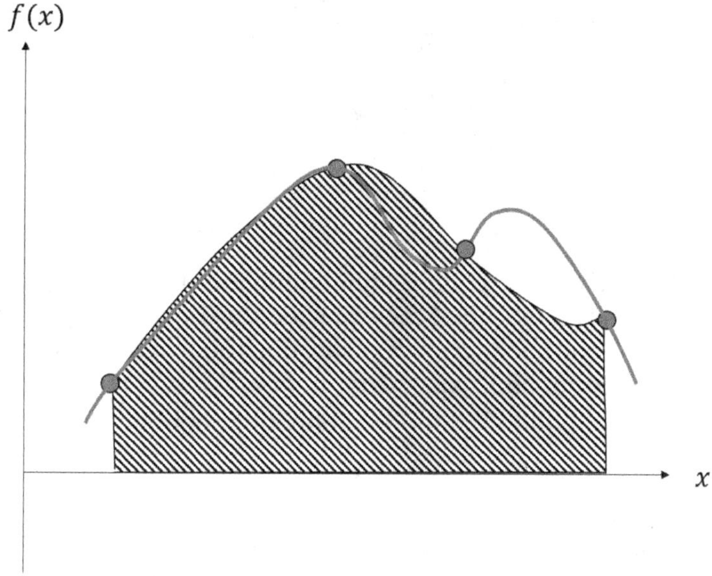

Figure 9.3 Graphical depiction of Simpson's 3/8 rule: It consists of taking the area undera cubic equation connecting four points.

This is the second Newton-Cotes closed integration formula. The label "1/3" stems from the fact that h is divided by 3 in Eq. (9.13). An alternative derivation is shown in Eq. 9.7 where the Newton-Gregory polynomial is integrated to obtain the same formula. Simpson's 1/3 rule can also be expressed using the format of Eq. (9.9):

$$I \cong (b-a)\,\frac{f(x_0)+4f(x_1)+f(x_2)}{6} \tag{9.14}$$

9.6.1 Derivation and Error Estimate of Simpson's 1/3 Rule

The derivation of Simpson's 1/3 rule, similar to the approach used in Eq. 9.6 for the trapezoidal rule, involves integrating the forward Newton-Gregory interpolating polynomial:

$$
\begin{aligned}
I = \int_{x_0}^{x_2} \Bigg[f(x_0) + \Delta f(x_0)\alpha + &\frac{\Delta^2 f(x_0)}{2}\alpha(\alpha-1) \\
&+ \frac{\Delta^3 f(x_0)}{6}\alpha(\alpha-1)(\alpha-2) \\
&+ \frac{f^4(\xi)}{24}\alpha(\alpha-1)(\alpha-2)(\alpha-3)h^4 \Bigg] dx
\end{aligned}
$$

It is worth noting that the polynomial has been written up to the fourth-order term instead of the expected third-order term. The rationale for this will become evident shortly. It is important to observe that the integration limits span from x_0 to x_2. Thus, after making the necessary simplifying substitutions, the integral ranges from $\alpha = 0$ to 2:

$$
\begin{aligned}
I = h\int_{0}^{2} \Bigg[f(x_0) + \Delta f(x_0)\alpha + &\frac{\Delta^2 f(x_0)}{2}\alpha(\alpha-1) \\
&+ \frac{\Delta^3 f(x_0)}{6}\alpha(\alpha-1)(\alpha-2) \\
&+ \frac{f^4(\xi)}{24}\alpha(\alpha-1)(\alpha-2)(\alpha-3)h^4 \Bigg] d\alpha
\end{aligned}
$$

which can be integrated to yield

$$
\begin{aligned}
I = h\Bigg[\alpha f(x_0) + + &\frac{\alpha^2 f(x_0)}{2}\alpha(\alpha-1) + \left(\frac{\alpha^3}{6} - \frac{\alpha^2}{4}\right)\Delta^2 f(x_0) \\
&+ \left(\frac{\alpha^4}{24} - \frac{\alpha^3}{6} + \frac{\alpha^2}{6}\right)\Delta^3 f(x_0) \\
&+ \left(\frac{\alpha^5}{120} - \frac{\alpha^4}{16} + \frac{11\alpha^3}{72} - \frac{\alpha^2}{8}\right) f^4(\xi)h^4 \Bigg]_0^2
\end{aligned}
$$

and evaluated for the limits to give

$$
I = h\left[2f(x_0) + \Delta f(x_0)\alpha + \frac{\Delta^2 f(x_0)}{3} + (0)\Delta^3 f(x_0) - \frac{1}{90} f^4(\xi)h^4 \right] \tag{9.15}
$$

It is noteworthy that the coefficient of the third divided difference yields a significant result of zero. As $\Delta f(x_0) = f(x_1) - f(x_0)$ and $\Delta^2 f(x_0) = f(x_2) - 2f(x_1) + f(x_0)$, Eq. 9.15 can be rewritten as

$$I = \frac{h}{3}[f(x_0) + 4f(x_1) + f(x_2)] - \frac{1}{90} f^4(\xi)h^5$$

The first term refers to Simpson's 1/3 rule, while the second term pertains to the truncation error. The omission of the third divided difference leads to a significant finding that the formula achieves third-order accuracy, where $a = x_0$, $b = x_2$, and $x_1 =$ the point midway between a and b, which is given by $(b + a)/2$.

It is worth noting that, as indicated by Equation (9.14), the middle point carries a weight of two-thirds, while the two end points carry a weight of one-sixth each.

It has been demonstrated that utilising Simpson's 1/3 rule on a single segment results in a truncation error of

$$E_t = -\frac{1}{90} f^4(\xi)h^5$$

Or because, $h = (b - a)/2$,

$$E_t = -\frac{(b - a)^5}{2880} f^4(\xi) \tag{9.16}$$

Where ξ is located within the interval from a to b. Therefore, it can be concluded that Simpson's 1/3 rule is more accurate compared to the trapezoidal rule. However, a comparison with Equation (9.10) suggests that it is more precise than anticipated. The error is proportional to the fourth derivative, instead of being proportional to the third derivative. This phenomenon occurs due to the integration of the interpolating polynomial, as demonstrated in 9.15. Specifically, the coefficient of the third-order term becomes zero. As a result, Simpson's 1/3 rule maintains a high level of accuracy despite being derived from a limited number of data points. In simple terms, it offers accurate solutions for cubic polynomials, even though it is derived from a parabola.

Example 9.6 Use Eq. (9.14) to integrate

$$f(x) = 0.2 + 25x - 200x^2 + 675x^3 - 900x^4 + 400x^5$$

from $a = 0$ to $b = 0.8$. Recall that the exact integral is 1.640533.

Solution. $f(0) = 0.2, \ f(0.4) = 2.456, \quad f(0.8) = 0.232$

Therefore, Eq. (9.14) can be used to compute

$$I \cong 0.8\frac{0.2 + 4(2.456) + 0.232}{6} = 1.367467$$

which represents an exact error of

$$E_t = 1.640533 - 1.367467 = 0.2730667$$
$$\varepsilon_t = 16.6\%$$

which is approximately 5 times more accurate than for a single application of the trapezoidal rule (Example 9.2),

The estimated error is

$$E_a = -\frac{(0.8)^5}{2880}(-2400) = 0.2730667$$

where -2400 is the average fourth derivative for the interval.
As was the case in Example 9.2, the error is approximate (E_a) because the average fourth derivative is not an exact estimate of $f^4(\xi)$. However, because this case deals with a fifth order polynomial, the result matches.

Example 9.7 Compute the approximate value of $\int_0^2 x^4\,dx$ by taking four sub-intervals and compare it with the exact value.

Solution. For four sub-intervals of $[0, 2]$, we have $h = \frac{2}{4} = \frac{1}{2} = 0.6$.
We tabulate $f(x) = x^4$

x	0	0.5	1	1.5	2
$f(x)$	0	0.0625	1	5.062	16

By trapezoidal rule, we get

$$\int_0^2 x^4\,dx \approx \frac{0.5}{2}[0 + 2 \times (0.0625 + 1.0 + 5.062) + 16.0]$$

$$\approx \frac{1}{4}[12.2690 + 16.0]$$

$$= \frac{28.2690}{4} = 7.0672$$

By Simpson's one-third rule, we get

$$\int_0^2 x^4\,dx \approx \frac{0.5}{3}[0 + 4 \times (0.0625 + 5.062) + 2 \times 1.0 + 16.0]$$

$$= \frac{1}{6}[4 \times 5.135 + 18.0] = \frac{38.5380}{6} = 6.4230$$

Exact value $= \frac{2^5}{5} = \frac{32}{5} = 6.4$

Error in the result by trapezoidal rule = 6.4 – 7.0672 =– 0.6672

Error in the result by Simpson's one third rule = 6.4 – 6.4230 = – 0.0230

Example 9.8 Evaluate the following integral:
$\int_0^1 (4x - 3x^2)dx$ by taking $n = 10$ and using the following rules:

(i) Trapezoidal rule and
(ii) Simpson's one-third rule. Also compare them with the exact value and find the error in each case.

Solution. We tabulate Numerical Integration $f(x) = 4x - 3x^2$, for $x = 0, 0.1, 0.2, \ldots, 1.0$.

x	0	0.1	0.2	0.3	0.4	0.5	0.6	0.7	0.8	0.9	1.0
$f(x)$	0	0.37	0.68	0.93	1.12	1.25	1.32	1.33	1.28	1.17	1.0

(i) Using trapezoidal rule, we have

$$\int_0^1 (4x - 3x^2)dx$$

$$= \frac{0.1}{2}[0 + 2(0.37 + 0.68 + 0.931.12 + 1.25 + 1.32 + 1.33 + 1.28 + 1.17)1.0]$$

(ii) Using Simpson's one-third rule, we have

$$\int_0^1 (4x - 3x^2)dx$$

$$= \frac{0.1}{3}[0 + 4(0.37 + 0.93 + 1.25 + 1.33 + 1.17) + 2(0.68 + 1.12 + 1.32 + 1.28)1.0]$$

$$= \frac{0.1}{3}[4 \times 5.05 + 2 \times 4.40 + 1.0] = \frac{0.1}{3} \times [30.0] = 1.00$$

(iii) Exact value $= 1.0$

Error in the result by trapezoidal rule is 0.005 and there is no error in the result by Simpson's one-third rule.

Example 9.9 Approximate the integral

$$I = \int_0^1 \frac{dx}{1+x}$$

Solution. The true value is $I = \log(2) \approx 0.693147$. Using the Simpson's rule, we get

$$I_s = \frac{1}{6}\left[1 + \frac{8}{3} + \frac{1}{2}\right] = \frac{25}{36} \approx 0.694444$$

Therefore, the error is $I - I_s \approx 0.001297$

9.7 Simpson's (3/8) Rule

Similar to the process used to derive the trapezoidal and Simpson's 1/3 rule, a third-order Lagrange polynomial can be fitted to four points and then integrated

$$I = \int_a^b f(x)dx \cong \int_a^b f_3(x)dx$$

to yield

$$I \cong \frac{3h}{8}[f(x_0) + 3f(x_1) + 3f(x_2) + f(x_3)]$$

Where $h = (b - a)/3$

The equation is usually referred to as Simpson's 3/8 rule due to the multiplication of h by 3/8. This formula represents the third closed integration method developed by Newton-Cotes. Another way to represent the 3/8 rule using Equation (9.14):

$$I \cong (b - a)\,\frac{f(x_0) + 3f(x_1) + 3f(x_2) + f(x_3)}{8} \tag{9.17}$$

Therefore, the weights given to the two interior points are three-eighths, while the end points have a weight of one-eighth. The error of Simpson's 3/8 rule is

$$E_t = -\frac{3}{80}f^4(\xi)h^5$$

Or because, $h = (b - a)/3$,

$$E_t = -\frac{(b - a)^5}{6480}f^4(\xi) \tag{9.18}$$

Due to the larger denominator of Eq. (9.18) compared to Eq. (9.16), the 3/8 rule is considered to be slightly more accurate than the 1/3 rule.

The preferred method is usually Simpson's 1/3 rule, which achieves third-order accuracy using only three points, as opposed to the four points needed for the 3/8 version. Nevertheless, the 3/8 rule proves useful in cases where the number of segments is odd. In Example 9.6, Simpson's rule was employed to integrate the function for four segments. Suppose you wanted to obtain an estimate for five segments. An alternative approach is to utilise a multi-application version of the trapezoidal rule. It may not be recommended due to the significant truncation error associated with this approach. A possible alternative is to use Simpson's 1/3 rule for the initial two segments and Simpson's 3/8 rule for the remaining three segments (Fig. 9.4). By following this approach, we can achieve a precise estimate that maintains a high level of accuracy throughout the entire interval.

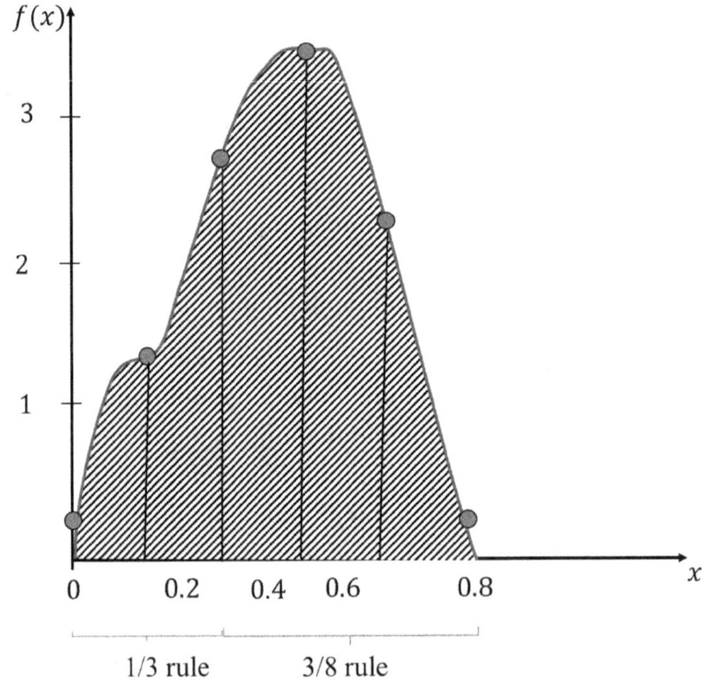

Figure 9.4 Illustration of how Simpson's 1/3 and 3/8 rules can be applied in tandem to handle multiple applications with odd numbers of intervals

Example 9.10

(a) Use Simpson's 3/8 rule to integrate

$$f(x) = 0.2 + 25x - 200x^2 + 675x^3 - 900x^4 + 400x^5$$

from $a = 0$ to $b = 0.8$.

(b) Use it in conjunction with Simpson's 1/3 rule to integrate the same function for five segments.

Solution.

(a) A single application of Simpson's 3y8 rule requires four equally spaced points:

$$f(0) = 0.2 \qquad f(0.2667) = 1.432724$$
$$f(0.5333) = 3.487177 \qquad f(0.8) = 0.232$$

Using Eq. (9.17),

$$I \cong 0.8\frac{0.2 + 3(1.432724 + 3.487177) + 0.232}{8} = 1.519170$$

$$E_t = 1.640533 - 1.519170 = 0.1213630 \qquad \varepsilon_t = 7.4\%$$

$$E_a = -\frac{(0.8)^5}{6480}(-2400) = 0.1213630$$

(b) The data needed for a five-segment application ($h = 0.16$) is

$$f(0) = 0.2 \qquad f(0.16) = 1.296919$$
$$f(0.32) = 1.743393 \qquad f(0.48) = 3.186015$$
$$f(0.64) = 3.181929 \qquad f(0.80) = 0.232$$

The integral for the first two segments is calculated using Simpson's 1/3 rule:

$$I \cong 0.32\frac{0.2 + 4(1.296919) + 1.743393}{6} = 0.3803237$$

For the last three segments, the 3/8 rule can be used to obtain

$$I \cong 0.48\frac{1.743393 + 3(3.186015 + 3.181929) + 0.232}{8} = 1.264754$$

The total integral is calculated by summing the two outcomes:

$$I = 0.3803237 + 1.264753 = 1.645077$$

$$E_t = 1.640533 - 1.645077 = -0.00454383 \qquad \varepsilon_t$$
$$= -0.28\%$$

Example 9.11 Find Solution of an equation $2x^3 - 4x + 1$ using Simpson's 3/8 rule $x_1 = 2$ and $x^2 = 4$, Step value $(h) = 0.5$

Solution. Equation is $f(x) = 2x^3 - 4x + 1$

The value of table for x and y

x	2	2.5	3	3.5	4
y	9	22.25	43	72.75	113

using Simpson's 3/8 rule,

$$\int ydx = \frac{3h}{8}[(y_0 + y_4) + 2(y_3) + 3(y_1 + y_2)]$$

$$\int ydx = \frac{3 \times 0.5}{8}[(9 + 113) + 2(72.75) + 3(22.25 + 43)]$$

$$\int ydx = \frac{3 \times 0.5}{8}[(9 + 113) + 2(72.75) + 3(62.25)]$$

$$\int ydx = 86.8594$$

Solution by Simpson's $\frac{3}{8}$ Rule is 86.8594 .

9.8 Newton-Cotes closed formulas of higher order

As previously mentioned, the trapezoidal rule and both of Simpson's rules belong to a group of integrating equations called the Newton-Cotes closed integration formulas. The formulas and their truncation-error estimates are presented in Table 9.2 for easy reference. It is worth noting that, similar to Simpson's 1/3 and 3/8 rules, the five and six-point formulas exhibit the same order error. The higher-point formulas exhibit this general characteristic, resulting in a preference for even-segment–odd-point formulas such as the 1/3 rule and Boole's rule. It is important to note that in engineering practice, higher-order formulas (greater than four-point) are rarely utilised. The use of Simpson's rules is generally adequate for a wide range of applications. Enhancing accuracy can be achieved by using the multi-application version.

Table 9.2 Newton-Cotes closed integration formulas

Segments (n)	Points	Name	Formula	Truncation Error
1	2	Trapezoidal rule	$(b-a)\frac{f(x_0)+f(x_1)}{2}$	$-\frac{1}{12}h^3 f''(\xi)$
2	3	Simpson's 1/3 rule	$(b-a)\frac{f(x_0)+4f(x_1)+f(x_2)}{6}$	$-\frac{1}{90}h^5 f^{(4)}(\xi)$
3	4	Simpson's 3/8 rule	$(b-a)\frac{f(x_0)+3f(x_1)+3f(x_2)+f(x_3)}{2}$	$-\frac{3}{80}h^5 f^{(4)}(\xi)$
4	5	Boole's rule	$(b-a)\frac{7f(x_0)+32f(x_1)+12f(x_2)+32f(x_3)+7f(x_4)}{2}$	$-\frac{8}{945}h^7 f^{(6)}(\xi)$
5	6		$(b-a)\frac{19f(x_0)+75f(x_1)+50f(x_2)+50f(x_3)+75f(x_4)+19f(x_5)}{2}$	$-\frac{275}{12096}h^7 f^{(6)}(\xi)$

In addition, for situations where precise and accurate results are necessary, methods like Romberg integration or Gauss quadrature can be considered as effective alternatives.

9.9 Gaussian Numerical Integration Formula

All of the derived rules so far can be expressed as follows:

$$I(f) = \int_a^b f(x)dx \approx w_0 f(x_0) + w_1 f(x_1) + \cdots + w_n f(x_n) \tag{9.19}$$

Here weights (w_i) are non-negative constants. The nodes are chosen so that, for polynomials of degree $\leq n$, the quadrature rule holds true exactly. The Newton-Conte formula of order n is used in reference to these procedures. However, by carefully selecting the nodes, it is possible to make such a rule exact for polynomials of degree $< 2n + 1$. This is how Gaussian rules work fundamentally. Let's consider one particular case

$$\int_{-1}^{1} f(x)dx \approx \sum_{i=0}^{n} w_i f(x_i) \tag{9.20}$$

The weights w_i and the nodes x_i $(i = 0,\cdots,n)$ are to be chosen in such a way that the error

$$E_n(f) = \int_{-1}^{1} f(x)dx - \sum_{i=0}^{n} w_i f(x_i) \tag{9.21}$$

is zero when f(x) is a polynomial of degree ≤ 2n + 1. To derive equations for the nodes and weights, we first note that

$$E_n(a_0 + a_1x + a_1x^2 + \cdots + a_mx^m) \\ = a_0E_n(1) + a_1E_n(x) + \cdots + a_nE_n(x^m)$$

Therefore, $E_n(f)$ for every polynomial of degree $\leq m$ if and only if $E_n(x^i) = 0$ for $\quad i = 0,1, \dots, m$.

Case 1: $n = 0$. Since there are two parameters, namely, w_0 and x_0, we consider the requiring $E_0(1) = E_0(x) = 0$. This gives $\int_{-1}^{1} 1dx - w_0 = 0$, and $\int_{-1}^{1} xdx - w_0\, x_0 = 0$. These gives $w_0 = 2$ and $x_0 = 0$. Thus, we have the formula

$$\int_{-1}^{1} f(x)dx \approx 2f(0) \tag{9.22}$$

which is the required Gaussian quadrature for $n = 0$.

Case 2: $n = 1$. There are two parameters, namely, w_0 , w_1 , x_0 and x_1 and thus we put four constraints on these parameters:

$$E_1(x^i) = \int_{-1}^{1} x^i dx - (w_0\, x_0^i + w_1 x_1^i) = 0\,, \qquad i = 0,1,2,3.$$

This gives a system of nonlinear equations

$w_1 + w_2 = 2\,,\ w_1x_1 + w_2x_2 = 0\,, w_1x_1^2 + w_2x_2^2 = \frac{2}{3}\,, w_1x_1^3 + w_2x_2^3 = 0$

The solutions are $w_1 = w_2 = 1$ and $x_1 = -1/\sqrt{3}$ and $x_2 = 1/\sqrt{3}$ which lead to the unique formula

$$\int_{-1}^{1} f(x)dx \approx f\left(-\frac{1}{\sqrt{3}}\right) + f\left(\frac{1}{\sqrt{3}}\right) = I_{G1}(f) \tag{9.23}$$

Case 3: General. There are $2(n+1)$ free parameters x_i and w_i for $i = 0,1,\cdots,n$. The equations to be solved are $E_1(x^i) = 0$, $i = 0,1,\ldots,2n+1$ or

$$\sum_{j=0}^{n} w_j x_j^i = \begin{cases} 0, & i = 1,3,\ldots,2n+1 \\ \dfrac{2}{i+1}, & i = 0,2,\ldots,2n \end{cases}$$

These are nonlinear equations, and the solution is not evident. Most computer software includes programs for creating nodes and weights, as well as doing numerical integration. The formulas shown above are known as the Gaussian numerical integration formula or Gaussian quadrature. Please keep in mind that this formula only applies to integrals across the range of -1 to 1. This limitation can be readily overcome by using a linear change of variable

$$x = \frac{b + a + t(b-a)}{2}, \qquad -1 \le t \le 1. \tag{9.24}$$

Thus, an integral $I(f) = \int_a^b f(x)dx$, can be transferred to

$$I(f) = \frac{b-a}{2} \int_{-1}^{1} f\left(\frac{b+a+t(b-a)}{2}\right) dt\,.$$

The following theorem provides the error formula for the Gaussian quadrature.

Example 9.12 Approximate the integral

$$I = \int_0^1 \frac{dx}{1+x}$$

Solution. Note that the true value is $I = \log(2) \approx 0.693147$.
To use the Gaussian quadrature, we first need to make the linear change of variable (9.24) with $a = 0$ and $b = 1$ and we get

$$x = \frac{t}{2}, \qquad -1 \leq t \leq 1 .$$

Thus, the required integration is

$$I = \int_0^1 \frac{dx}{1+x} = \int_{-1}^1 \frac{dt}{3+t}$$

We need to take $f(t) = 1/(3+t)$ in the Gaussian quadrature formula (9.23) and we get

$$\int_0^1 \frac{dx}{1+x} = \int_{-1}^1 \frac{dt}{3+t} \approx f\left(-\frac{1}{\sqrt{3}}\right) + f\left(\frac{1}{\sqrt{3}}\right) \approx 0.692308 \approx I_{G1}(f)$$

Therefore, the error is $I - I_{G1} \approx 0.000839$.

9.10 Multiple Integrals

Multiple integrals are often used in engineering. A general equation for calculating the average of a two-dimensional function can be expressed as:

$$\overline{f} = \frac{\int_c^d \left(\int_a^b f(x,y)dx\right) dy}{(d-c)(b-a)} \tag{9.25}$$

The numerator is called a double integral.

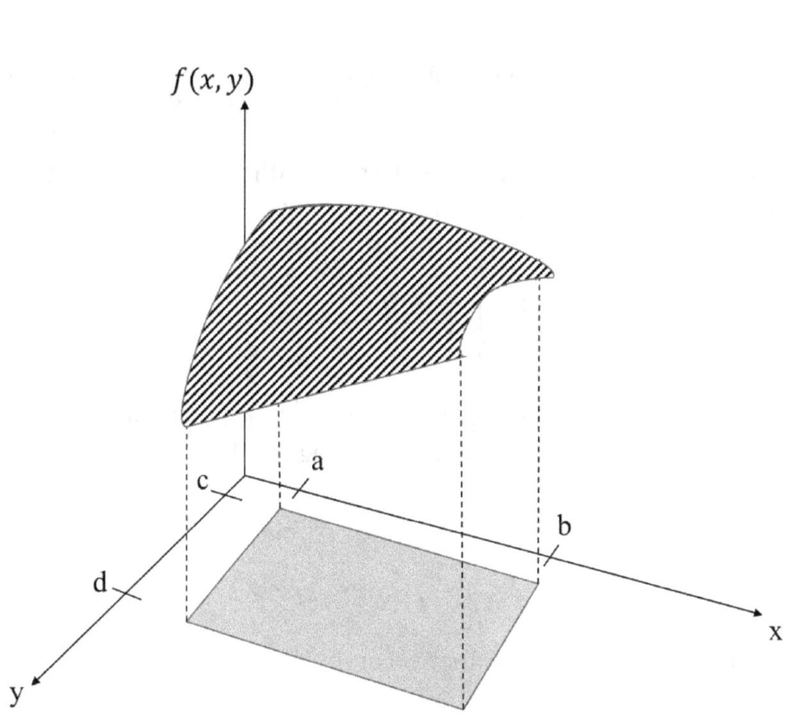

Figure 9.5 Double integral as the area under the function surface.

The procedures can be easily used to assess various integrals. An illustrative instance involves computing the double integral of a function across a rectangular region (Fig. 9-5).

As a reminder from calculus, these integrals can be calculated as iterated integrals.

$$\int_c^d \left(\int_a^b f(x,y)dx \right) dy = \int_a^b \left(\int_c^d f(x,y)dy \right) dx \tag{9.26}$$

Therefore, the integral in one of the dimensions is assessed first. The outcome of this initial integration is incorporated into the second dimension. Equation (9.26) asserts that the sequence of integration does not have any significance.

A numerical double integral is derived from the same concept. Initially, techniques such as the multiple-segment trapezoidal or Simpson's rule would be utilised in the primary dimension while keeping each value of the secondary dimension constant. Subsequently, the approach would be used to incorporate the second dimension.

Example 9.13 Use Gaussian quadrature with $n = 3$ in both dimensions to approximate the integral

$$\int_{1.4}^{2.0}\int_{1.0}^{1.5} \ln(x+2y)dydx.$$

Solution. Before employing Gaussian quadrature to approximate this integral, we need to transform the region of integration

$$R = \{(x,y)|1.4 \leq x \leq 2.0\,, 1.0 \leq y \leq 1.5\}$$

Into $\hat{R} = \{(u,v)|-1 \leq u \leq 1, -1 \leq v \leq 1\}$.

The linear transformations that accomplish this are

$u = \frac{1}{2.0-1.4}(2x - 1.4 - 2.0)$ and $v = \frac{1}{1..5-1.0}(2y - 1.0 - 1.5)$, or, equivalently, $x = 0.3u + 1.7$ and $y = 0.25v + 1.25$. Employing this change of variables gives an integral on which Gaussian quadrature can be applied:

$$\int_{1.4}^{2.0}\int_{1.0}^{1.5} \ln(x+2y)dydx = 0.075\int_{-1}^{1}\int_{-1}^{1} \ln(0.3u+0.5v+4.2)dudv.$$

The Gaussian quadrature formula for $n = 3$ in both u and v requires that we use the nodes

$u_1 = v_1 = r_{3,2} = 0\,, u_0 = v_0 = r_{3,1} = -0.7745966692\,,$
and $u_2 = v_2 = r_{3,3} = 0.7745966692$

The associated weights are $c_{3,2} = 0.\overline{8}$ and $c_{3,1} = c_{3,3} = 0.\overline{5}$. The resulting approximation is

$$\int_{1.4}^{2.0}\int_{1.0}^{1.5} \ln(x+2y)dydx$$

$$= 0.075\sum_{i=1}^{3}\sum_{j=1}^{3} c_{3,i}\,, c_{3,j}\ln\left(0.3r_{3,i}+0.5r_{3,j}+4.2\right)$$

$$= 0.4295545313$$

9.11 Exercises

1. Evaluate the following integral:

$$\int_0^{\pi/2} (6+3\cos x)dx$$

(a) analytically; (b) Trapezoidal rule;
(d) Simpson's 1/3 rule; (e) Simpson's 3/8 rule

2. Evaluate the following integral:

$$\int_0^3 (1-e^{-2x})dx$$

(a) analytically; (b) Trapezoidal rule;
(d) Simpson's 1/3 rule; (e) Simpson's 3/8 rule

3. Integrate the following function both analytically and using Simpson's rules, with $n = 4$ and 5. Discuss the results.

$$\int_{-3}^{5} (4x-3)^3 dx$$

4. Write down the errors in the approximation of

$$\int_0^1 x^4 dx \text{ and } \int_0^1 x^5 dx$$

by the Trapezoidal rule and Simpson's rule. Hence find the value of the constant C for which the Trapezoidal rule gives the exact result for the calculation of $\int_0^1 (x^5 - Cx^4)dx$.

5. Estimate the effect of data inaccuracy on results computed by Trapezoidal and Simpson's rule.

6. Determine the coefficients in the quadrature formula

$$\int_0^{2h} x^{-\frac{1}{2}} f(x)dx = (2h)^{\frac{1}{2}}\left(w_0 f(0) + w_1 f(h) + w_2 f(2h)\right)$$

7. Use the two-point Gaussian quadrature rule to approximate

$$\int_{-1}^{1} \frac{dx}{x+2}$$

and compare the result with the trapezoidal and Simpson rules.

8. Apply Trapezoidal, Simpson and Gaussian methods to evaluate

a) $I = \int_0^{\pi/2} \frac{\cos x}{1+\cos^2 x} dx$ (exact value ≈ 0.623225)

b) $I = \int_0^{\pi} \frac{1}{5+4\cos x} dx$ (exact value ≈ 1.047198)

c) $I = \int_0^1 e^{-x^2} dx$ (exact value ≈ 0.746824)

d) $I = \int_0^{\pi} \sin^3 x \cos^4 x \, dx$ (exact value ≈ 0.114286)

e) $I = \int_0^1 (1 + e^{-x^2} \sin(4x)) dx$ (exact value ≈ 1.308250)

9. Solve

$$f(x) - \int_0^1 (x+t) f(t) dt = \frac{3}{2} x - \frac{5}{6}$$

10. Use $n = m = 4$ to approximate the following double integrals, and compare the results to the exact answers.

$$\int_{2.1}^{2.5} \int_{1.2}^{1.4} xy^2 dy dx$$

11. Use $n = m = 4$ to approximate the following double integrals, and compare the results to the exact answers.

$$\int_{0}^{0.5} \int_{0}^{0.5} e^{y-x} dy dx$$

12. Approximate the following integrals using Gaussian quadrature with $n = 2$, and compare your results to the exact values of the integrals.

a) $\int_1^{1.5} x^2 \ln x \, dx$

b) $\int_1^{1.5} x^2 e^{-x} dx$

c) $\int_0^{0.35} \frac{2}{x^2-4} dx$

d) $\int_0^{\pi/4} x^2 \sin x \, dx$

e) $\int_0^{\pi/4} e^{3x} \sin 2x \, dx$

13. Repeat Exercise 12 with $n = 3$.

14. Repeat Exercise 12 with $n = 4$.

15. Repeat Exercise 12 with $n = 5$.

16. Compute the Simpson's rule approximations $S(a, b)$, $S(a, (a + b)/2)$, and $S((a + b)/2, b)$ for the following integrals, and verify the estimate given in the approximation formula.

a) $\int_1^{1.5} x^2 \ln x \, dx$

b) $\int_0^1 x^2 e^{-x} dx$

c) $\int_0^{0.35} \frac{2}{x^2-4} dx$

d) $\int_1^{1.6} \frac{2x}{x^2-4} dx$

e) $\int_3^{3.5} \frac{x}{\sqrt{x^2-4}} dx$

17. Approximate the following integrals using the Trapezoidal rule.

a) $\int_{0.5}^1 x^4 dx$

b) $\int_0^{0.5} \frac{2}{x-4} dx$

c) $\int_0^{0.35} \frac{2}{x^2-4} dx$

d) $\int_1^{1.6} \frac{2x}{x^2-4} dx$

e) $\int_0^{\pi/4} x \sin x \, dx$

18. Approximate the following integrals using the Trapezoidal rule.

a) $\int_{-0.25}^{0.25} (\cos x)^2 dx$

b) $\int_{0.75}^{1.3} ((\sin x)^2 - 2x \sin x + 1) dx$

c) $\int_e^{e+1} \frac{1}{x \ln x} dx$

d) $\int_{-0.5}^0 x \ln(x + 1) \, dx$

19. Repeat Exercise 17 using Simpson's rule.

20. Repeat Exercise 18 using Simpson's rule.

21. The Trapezoidal rule applied to $\int_0^2 f(x) dx$ gives the value 4, and Simpson's rule gives the value 2. What is $f(1)$?

22. Given the function f at the following values,

x	1.8	2	2.2	2.4	2.6
$f(x)$	3.12014	4.42569	6.04241	8.03014	10.46675

approximate $\int_{1.8}^{2.6} f(x)dx$ using all the appropriate quadrature formulas of this section.

CHAPTER 10

Numerical Methods for Solving Systems of Equations

10.1 Introduction to Linear Algebra

Engineering and applied sciences frequently require solving systems of linear algebraic equations and calculating eigenvalues and eigenvectors of matrices. The readers are assumed to have knowledge of the theory of determinants and elements of matrix algebra, as these concepts offer a convenient means of representing linear algebraic equations. We are going to address systems of equations since most interesting problems include more than one variable. Prior to discussing nonlinear systems, it is essential to thoroughly examine algorithms for solving linear systems due to two primary reasons. The linear case is necessary for the nonlinear case. We will simplify the solution of nonlinear problems by solving linear systems of equations. Furthermore, this enables us to incorporate concepts from linear algebra that are essential for future advancements.

Consider the system of n linear equations and n unknowns, is defined as follows:

$$\begin{aligned} a_{11}x_1 + a_{12}x_2 + \cdots + a_{1n}x_n &= b_1\,, \\ a_{21}x_1 + a_{22}x_2 + \cdots + a_{2n}x_n &= b_2\,, \\ \vdots \qquad\qquad\qquad &\quad \vdots \\ a_{n1}x_1 + a_{n2}x_2 + \cdots + a_{nn}x_n &= b_n\,. \end{aligned} \tag{10.1}$$

where the a's are constant coefficients and the b's are constants. The system can be represented by the matrix equation

$$Ax = b\,,$$

With

$$A = \begin{pmatrix} a_{11} & a_{12} & \cdots & a_{1n} \\ a_{21} & a_{22} & \cdots & a_{2n} \\ \vdots & \vdots & \ddots & \vdots \\ a_{n1} & a_{n2} & \cdots & a_{nn} \end{pmatrix}, \quad x = \begin{pmatrix} x_1 \\ x_2 \\ \vdots \\ x_n \end{pmatrix},$$

$$b = \begin{pmatrix} b_1 \\ b_2 \\ \vdots \\ b_n \end{pmatrix}.$$

Theorem 10.1. Let n be a positive integer, and let A be given as in (10.1). Then the following statements are equivalent

- $\det(A) \neq 0$
- For each right hand side *b,* the system (10.1) has unique solution x.
- For $b = 0$, the only solution for the system (10.1) is the zero solution.

This chapter presents a formalised approach for solving systems of equations through elimination. The basic elimination method, known as Gauss, was already in use prior to Gauss's time. The fangcheng methods were established in China several centuries prior to the emergence of Gauss. The method is currently included in the middle-school curriculum. Therefore, we will briefly review it in order to establish practical notation for deriving its fundamental properties.

10.2 Direct Methods

The linear equation $Ax = b$, is solved when A is a linear matrix and b is a vector. These issues are common in various fields of study, including science, engineering, economics, and finance. There is no universally superior technique for all situations.

There are two main categories of methods: **iterative methods** and **direct methods**. If round-off error is not present, these methods would produce the precise solution within a limited number of iterations. The primary direct method is Gaussian elimination, which primarily focuses on the matrix A and involves decomposing it into a product $A = LU$. We will concentrate on the situation in which A is triangular. Direct methods aim to quickly and easily solve the system in n^2 steps by creating a triangular decomposition of matrix A. This is achieved through a process called back-substitution, which will be described in detail.

Direct methods, which involve computing factorizations, are computationally expensive to solve on a desktop machine when A contains a large number of rows, typically tens or hundreds of thousands. This issue may persist indefinitely, necessitating the termination of the Terminal window to prevent computer freezing. However, direct methods are a convenient and effective approach for smaller matrices.

10.3 Gaussian Elimination

Gaussian elimination is the formal term used to describe the process of solving systems of linear equations by systematically eliminating unknowns and reducing them to systems of lower order. The approach is usually covered in high school algebra or undergraduate linear algebra courses, generally used to transform a matrix into its row-echelon form. Gaussian elimination is defined precisely, which is essential for its implementation on a computer and for analyzing the impact of rounding errors that arise during computation.

Let us now discuss the Gaussian Elimination method for $n = 3$. The approach for a general $n \times n$ system is similar. Let us consider the 3×3 system

$$\begin{aligned} a_{11}x_1 + a_{12}x_2 + a_{13}x_3 &= b_1 \qquad (E1) \\ a_{21}x_1 + a_{22}x_2 + a_{23}x_3 &= b_2 \qquad (E2) \\ a_{31}x_1 + a_{32}x_2 + a_{33}x_3 &= b_3 \qquad (E3) \end{aligned} \tag{10.2}$$

Step 1: Let's assume that $a_{11} \neq 0$ (otherwise interchange the row for which the coefficient of x_1 is non-zero).

Now let us eliminate x_1 from $(E2)$ and $(E3)$. To clarify,

$$m_{21} = \frac{a_{21}}{a_{11}}, \quad m_{31} = \frac{a_{31}}{a_{11}}$$

To obtain the result, multiply $(E1)$ by m_{21} and subtract it from $(E2)$. Additionally, multiply $(E1)$ by m_{31} and subtract it from $(E3)$.

$$\begin{aligned} a_{11}x_1 + a_{12}x_2 + a_{13}x_3 &= b_1 && (E1) \\ a_{22}^{(2)}x_2 + a_{23}^{(2)}x_3 &= b_2^{(2)} && (E2) \\ a_{32}^{(2)}x_2 + a_{33}^{(2)}x_3 &= b_3^{(2)} && (E3) \end{aligned}$$

The coefficients $a_{ij}^{(2)}$ are defined by

$$\begin{aligned} a_{ij}^{(2)} &= a_{ij} - m_{1j}a_{1j}\,, && i,j = 2,3 \\ b_i^{(2)} &= b_i - m_{i1}b_1\,, && i = 2,3 \end{aligned}$$

Step 2: Let $a_{22}^{(2)} \neq 0$, and remove x_2 from equation $(E3)$. Define the concept,

$$m_{32} = \frac{a_{32}^{(2)}}{a_{22}^{(2)}}$$

Subtract m_{32} times $(E2)$ from $(E3)$ to obtain,

$$\begin{aligned} a_{11}x_1 + a_{12}x_2 + a_{13}x_3 &= b_1 && (E1) \\ a_{22}^{(2)}x_2 + a_{23}^{(2)}x_3 &= b_2^{(2)} && (E2) \\ a_{33}^{(3)}x_3 &= b_3^{(3)} && (E3) \end{aligned}$$

The new defined coefficients are as follows:

$$a_{33}^{(3)} = a_{33}^{(2)} - m_{32}a_{23}^{(2)}\,, \qquad b_3^{(3)} = b_3^{(2)} - m_{32}b_2^{(2)}$$

Step 3: Using back substitution to solve successively for x_3, x_2 and x_1 , we obtain,

$$x_3 = \frac{b_3^{(3)}}{a_{33}^{(3)}}$$

$$x_2 = \frac{b_2^{(2)} - a_{23}^{(2)}x_3}{a_{22}^{(2)}}$$

$$x_1 = \frac{b_1 - a_{12}x_2 - a_{13}x_3}{a_{11}} \tag{10.3}$$

The algorithm for $n = 3$ can be easily extended to a general $n \times n$ non-singular linear system.

The Gaussian elimination method is a precise and direct approach for solving linear systems. However, occasionally this method fails to provide the accurate solution, as demonstrated in the following example.

Example 10.1 When we solve the linear system

$$\begin{gathered} 6x_1 + 2x_2 + 2x_n = -2 \\ 2x_1 + \frac{2}{3}x_2 + \frac{1}{3}x_n = 1 \\ x_1 + 2x_2 - x_n = 0 \end{gathered}$$

This problem can be solved by a computer by applying the Gaussian elimination approach. The floating-point representation in the mantissa will have four digits, and all operations will be rounded.

The provided system is

$$\begin{gathered} 6.000x_1 + 2\,.000x_2 + 2.000x_n = -2.000 \\ 2.000x_1 + 0.667x_2 + 0.3333x_n = 1.000 \\ 1.000x_1 + 2.000x_2 - 1.000x_n = 0.000 \end{gathered}$$

After eliminating x_1 from the second and third equations, we get (with $m_{21} = 0.3333, m_{31} = 0.1667$)

$$\begin{gathered} 6.000x_1 + 2\,.000x_2 + 2.000x_n = -2.000 \\ 0.000x_1 + 0.0001x_2 + 0.3333x_n = 1.667 \\ 0.000x_1 + 1.667x_2 - 1.333x_n = 0.3334 \end{gathered}$$

After eliminating x_2 from the third equation, we get (with $m_{32} = 16670$)

$$\begin{gathered} 6.000x_1 + 2\,.000x_2 + 2.000x_n = -2.000 \\ 0.000x_1 + 1.667x_2 + 1.333x_n = 0.3334 \\ 0.000x_1 + 0.0000x_2 - 0.3332x_n = 1.667 \end{gathered}$$

with back substitution, we get the approximate solution as $x_1 = 2.602$, $x_2 = -3.801$ and $Dx_3 = -5.003$.

Example 10.2 Solve the linear system

$$\begin{aligned} x_1 + 2x_2 + x_3 &= 0 \\ 2x_1 + 2x_2 + 3x_3 &= 3 \\ -x_1 - 3x_2 \quad &= 2 \end{aligned}$$

To simplify the notation, it is important to remember that the variables x_1, x_2, and x_3 are not included in the algorithm until the last step.

Thus, the augmented matrix is used to represent the preceding linear system.

$$[A|b] = \left[\begin{array}{ccc|c} 1 & 2 & 1 & 0 \\ 2 & 2 & 3 & 3 \\ -1 & -3 & 0 & 2 \end{array}\right]$$

The augmented matrix gets row operations, and the unknowns are provided in the final step. The diagram displays multipliers positioned adjacent to arrows, representing the corresponding changes they cause:

$$\left[\begin{array}{ccc|c} 1 & 2 & 1 & 0 \\ 2 & 2 & 3 & 3 \\ -1 & -3 & 0 & 2 \end{array}\right] \xrightarrow[m_{31}=-1]{m_{21}=2} \left[\begin{array}{ccc|c} 1 & 2 & 1 & 0 \\ 2 & 2 & 3 & 3 \\ -1 & -3 & 0 & 2 \end{array}\right]$$

$$\downarrow m_{31} = \frac{1}{2}$$

$$[U|g] = \left[\begin{array}{ccc|c} 1 & 2 & 1 & 0 \\ 0 & -2 & 1 & 3 \\ 0 & 0 & \frac{1}{2} & \frac{1}{2} \end{array}\right]$$

Solving $U_x = g$,

$$x_3 = 1, \qquad x_2 = -1, \qquad x_1 = 1$$

Example 10.3 Solve the following system by Gauss elimination method:

$$\begin{aligned} 2x_1 - 2x_2 + 3x_3 &= 2 \\ x_1 + 2x_2 - x_3 &= 3 \end{aligned}$$

$$3x_1 - x_2 + 2x_3 = 1$$

Solution.

$$[A|b] = \left[\begin{array}{ccc|c} 2 & -2 & 3 & 2 \\ 1 & 2 & -1 & 3 \\ 3 & -1 & 2 & 1 \end{array}\right]$$

Transforming augmented matrix to echelon form, we obtain

$$\left[\begin{array}{ccc|c} 2 & -2 & 3 & 2 \\ 1 & 2 & -1 & 3 \\ 3 & -1 & 2 & 1 \end{array}\right] \xrightarrow{R_1 \leftrightarrow R_2} \left[\begin{array}{ccc|c} 1 & 2 & -1 & 3 \\ 2 & -2 & 3 & 2 \\ 3 & -1 & 2 & 1 \end{array}\right]$$

$$\xrightarrow[R_3 \to R_3 - 3R_1]{R_2 \to R_2 - 2R_1} \left[\begin{array}{ccc|c} 1 & 2 & -1 & 3 \\ 0 & -6 & 5 & -4 \\ 0 & -7 & 5 & -8 \end{array}\right]$$

$$\xrightarrow{R_3 \to R_3 - R_2} \left[\begin{array}{ccc|c} 1 & 2 & -1 & 3 \\ 0 & -6 & 5 & -4 \\ 0 & -1 & 0 & -4 \end{array}\right]$$

The given system of equations using the row-echelon form reduces to

$$\begin{gathered} x_1 + 2x_2 - x_3 = 3 \\ -6x_2 + 5x_3 = -4 \\ -x_2 = -4 \\ \therefore x_2 = 4 \end{gathered}$$

By solving,

$$x_1 = -1, \qquad x_2 = 4, \qquad x_3 = 4$$

10.4 Operations Count

The duration of calculations and the resulting round-off error are influenced by the number of floating-point arithmetic operations necessary to solve a routine problem. Multiplication and division operations on a computer typically take approximately the same amount of time, but they are significantly longer than addition or subtraction operations. The variations in execution time are contingent upon the specific computing system. The steps needed to solve a common linear system of n equations in n unknowns will be counted in order to illustrate the counting procedures for a specific method.

Since it's critical to understand the duration of a computation, we count the number of arithmetic operations required for Gaussian elimination. Let us divide the count into three segments.

The elimination step. We are currently tallying the number of additions/subtractions, multiplications, and divisions required to transition from the original system to the triangle system.

Step	Additions/ Subtractions	Multiplications	Divisions
1	$(n-1)^2$	$(n-1)^2$	$n-1$
2	$(n-2)^2$	$(n-2)^2$	$n-2$
.	.	.	.
.	.	.	.
.	.	.	.
$n-1$	1	1	1
Total	$\frac{n(n-1)(2n-1)}{6}$	$\frac{n(n-1)(2n-1)}{6}$	$\frac{n(n-1)}{2}$

Here we use the formula

$$\sum_{j=1}^{p} j = \frac{p(p+1)}{2}, \sum_{j=1}^{p} j^2 = \frac{p(p+1)(2p+1)}{6}, p \geq 1$$

Modification of the right side. Proceeding as previously, we obtain

Addition/Subtraction $= (n-1) + (n-2) + \cdots + 1 = \frac{n(n-1)}{2}$

Multiplication/Division $= (n-1) + (n-2) + \cdots + 1 = \frac{n(n-1)}{2}$

The back substitution.

Addition/Subtraction $= (n-1) + (n-2) + \cdots + 1 = \frac{n(n-1)}{2}$

Multiplication/Division $= n + (n-1) + \cdots + 1 = \frac{n(n-1)}{2}$

So total number of operations in obtaining x is

Addition/Subtraction

$$= \frac{n(n-1)(2n-1)}{6} + \frac{n(n-1)}{2} + \frac{n(n-1)}{2} = \frac{n(n-1)(2n+5)}{6}$$

Multiplication/Division

$$= \frac{n(n^2+3n-1)}{3}$$

When considering only multiplication and division, it is observed that for a large value of n, the number of operations needed for Gaussian elimination is approximately $\frac{1}{3}n^3$. As the value of n increases, the cost of solving the linear system also increases by a factor of 8. Furthermore, the primary expense associated with Gaussian elimination lies in the elimination process. To eliminate, we have Multiplication/Division

$$= \frac{n(n-1)(2n-1)}{6} + \frac{n(n-1)}{2} = \frac{1}{3}(n^3 - n) = \frac{1}{3}n^3\left(1 - \frac{1}{n^2}\right) \approx \frac{1}{3}n^3,$$

whereas the remaining steps counts only

Multiplication/Division

$$= \frac{n(n-1)}{2} + \frac{n(n+1)}{2} = n^2$$

Therefore, once the elimination part is completed, it is much less expensive to solve the linear system.

10.5 Pivoting

Pivoting refers to the process of rearranging the unknown variables within a system (10.1). The numerical representation of equations and unknowns does not impact the solutions, but it can influence the results of the elimination algorithm. If the pivot is zero, the entire process fails. Specifically, there may be a stage where the elimination algorithm fails due to the undefined multipliers. Algorithm 10.1 required a row interchange when one of the pivot elements $a_{kk}^{(k)} \neq 0$. The form of this row interchange is $(E_k) \leftrightarrow (E_p)$, where p is the smallest integer greater than k with $a_{pk}^{(k)} \neq 0$. Row interchanges are frequently required to minimise round-off error, even in cases where the pivot elements are not zero. One option is to either relocate the necessary row and column to their correct positions or implement a new numbering scheme for tracking purposes. Pivoting can be used to apply row and/or column exchanges to the original matrix, denoted as $\hat{A}$. In this case, Gaussian elimination with pivoting is simply the original elimination algorithm applied to $\hat{A}$ without any pivoting. If $a_{kk}^{(k)}$ is small in magnitude compared to $a_{jk}^{(k)}$, then the magnitude of the multiplier

$$m_{jk} = \frac{a_{jk}^{(k)}}{a_{kk}^{(k)}} \tag{10.4}$$

will have a magnitude greater than 1. A round-off error occurs when one of the terms $a_{kl}^{(k)}$ is multiplied by m_{jk} during the computation of $a_{jl}^{(k+1)}$, which causes the initial error. When performing backward substitution,

$$x_k = \frac{a_{k,n+1}^{(k)} - \sum_{j=k+1}^{n} a_{kj}^{(k)}}{a_{kk}^{(k)}} \tag{10.5}$$

When the value of $a_{kk}^{(k)}$ is small, any error in the numerator can be significantly increased due to the division by $a_{kk}^{(k)}$. In the following example, we will observe that round-off error can significantly impact calculations, even for small systems.

Example 10.4 Apply Gaussian elimination to the system

$$E_1: 0.003000x_1 + 59.14x_2 = 59.17$$
$$E_2: 5.291x_1 - 6.130x_2 = 46.78$$

using four-digit arithmetic with rounding and compare the results to the exact solution $x_1 = 10.00$ and $x_2 = 1.000$.

Solution. The first pivot element, $a_{11}^{(1)} = 0.003000$, is small, and its associated multiplier,

$$m_{21} = \frac{5.291}{0.003000} = 1763.6\bar{6}$$

rounds to the large number 1764. Performing $(E_2) - m_{21}(E_1) \rightarrow (E_2)$ and the appropriate rounding gives the system

$$0.003000x_1 + 59.14x_2 \approx 59.17$$
$$-104300x_2 \approx -104400\,.$$

instead of the exact system, which is

$$0.003000x_1 + 59.14x_2 \approx 59.17$$
$$-104309.37\bar{6}x_2 \approx -104309.37\bar{6}\,.$$

The disparity in the magnitudes of $m_{21}a_{13}$ and a_{23} has introduced round-off error, but the round-off error has not yet been propagated. Backward substitution yields

$$x_2 \approx 1.001\,,$$

which is a close approximation to the actual value, $x_2 \approx 1.001$. However, because of the small pivot $a_{11} = 0.003000$,

$$x_1 \approx \frac{59.17 - (59.14)(1.001)}{0.003000} = -10.00$$

contains the small error of 0.001 multiplied by $\frac{59.14}{0.003000} \approx 20000$.

This damages the estimation of the true value of x_1, which is 10.00.

This example is clearly contrived, and Figure 10-1 illustrates the convenience with which the error can occur. For larger systems, it is challenging to predict the occurrence of devastating round-off errors in advance.

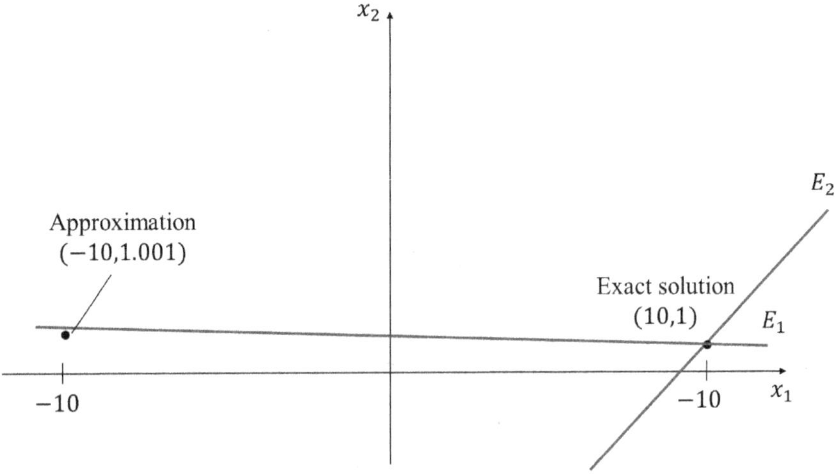

Figure 10.1 The approximation and the exact solution

Example 10.5 Use Gaussian elimination to solve the system

$$2x + y + z = 10$$
$$3x + 2y + 3z = 18$$
$$x + 4y + 9z = 16$$

Solution. Let us first eliminate x from the second and third equations. For this we multiply the first equation by $(-\frac{3}{2})$ and add it to the second to get

$$y + 3z = 6\,. \tag{i}$$

Similarly, we multiply the first equation by $\left(-\frac{1}{2}\right)$ and add it to the second to get

$$7y + 17z = 22 \tag{ii}$$

By solving, we get

$$-4z = -20 \text{ or } z = 5\,.$$

The upper-triangular form is therefore given by

$$2x + y + z = 10$$
$$y + 3z = 6$$
$$z = 5\,.$$

It follows that the required solution is $x = 7\,, y = -0\,, z = 5\,.$
The next example demonstrates the necessity of pivoting in the elimination method.

Example 10.6 Solve the system

$$0.0003120x_1 + 0.006032x_2 = 0.003328$$
$$0.5000x_1 + 0.8942x_2 = 0.9471$$

The exact solution is $x_1 = 1$ and $x_2 = 0.5\,.$

We first solve the system with pivoting. Let's write the system as

$$0.5000x_1 + 0.8942x_2 = 0.9471$$
$$0.000312x_1 + 0.006032x_2 = 0.003328$$

Using Gaussian elimination, the above system reduces to

$$0.5000x_1 + 0.8942x_2 = 0.9471$$
$$0.005474x_2 = 0.002737$$

Back substitution gives: $x_2 = 0.5$ and $x_1 = 1.0\,.$

Without pivoting, Gaussian elimination gives the system

$$0.000312x_1 + 0.006032x_2 = 0.003328$$
$$-8.7725x_2 = -5.3300$$

The back substitution gives

$$x_2 = 0.6076 \text{ and } x_1 = -1.0803$$

The effect of pivoting is clearly seen.

10.5.1 Partial Pivoting

Example 10.4 shows the difficulties that can occur when the pivot element $a_{kk}^{(k)}$ is relatively small compared to the entries $a_{ij}^{(k)}$, given that $k \le i \le n$ and $k \le j \le n$. The problem is addressed by performing pivoting, which involves selecting an element $a_{pq}^{(k)}$ with a larger magnitude as the pivot and replacing the kth and pth rows. If required, the kth and qth columns can be interchanged.

The most straightforward approach involves choosing an element in the same column that is positioned below the diagonal and possesses the highest absolute value. Specifically, we determine the smallest $p \geq k$ such that

$$\left|a_{pk}^{(k)}\right| = \max_{k \leq i \leq n} \left|a_{ik}^{(k)}\right|$$

and perform $(E_k) \leftrightarrow (E_p)$. No column interchange is used in this case.

Example 10.7 Apply Gaussian elimination to the system

$$\begin{aligned} E_1\colon\ 0.003000x_1 + 59.14x_2 &= 59.17 \\ E_2\colon\ 5.291x_1 - 6.130x_2 &= 46.78 \end{aligned}$$

using partial pivoting and four-digit arithmetic with rounding and compare the results to the exact solution $x_1 = 10.00$ and $x_2 = 1.000$.

Solution. The partial-pivoting procedure first requires finding

$$\max\left\{\left|a_{11}^{(1)}\right|, \left|a_{21}^{(1)}\right|\right\} = \max\{|0.003000|, |5.291|\} = |5.291| = \left|a_{21}^{(1)}\right|.$$

This requires that the operation $(E_2) \leftrightarrow (E_1)$ be performed to produce the equivalent system

$$\begin{aligned} E_1\colon\ 5.291x_1 - 6.130x_2 &= 46.78 \\ E_2\colon\ 0.003000x_1 + 59.14x_2 &= 59.17 \end{aligned}$$

The multiplier for this system is

$$m_{21} = \frac{a_{21}^{(1)}}{a_{11}^{(1)}} = 0.0005670\,.$$

and the operation $(E_2 - m_{21}E_1) \leftrightarrow (E_2)$ reduces the system to

$$5.291x_1 - 6.130x_2 \approx 46.78$$
$$59.14x_2 \approx 59.14\,.$$

The four-digit answers resulting from the backward substitution are the correct values
$x_1 = 10.00$ and $x_2 = 1.000$.

The technique just described is called partial pivoting (or maximal column pivoting).

10.6 Gauss-Jordan Method

The primary difference between this and the Gauss elimination method is that when an unknown is eliminated, it is removed from all of the equations. The solution can be obtained without the need for back substitution, as showed by the following example.

Example 10.8 Solve the system (see Example 7.5)

$$2x + y + z = 10$$
$$3x + 2y + 3z = 18$$
$$x + 4y + 9z = 16$$

by the Gauss-Jordan method.

Solution. Elimination of x from the second and third equations is done as in 'Gauss elimination' and we obtain the system

$$2x + y + z = 10$$
$$\left(\frac{1}{2}\right)y + \left(\frac{3}{2}\right)z = 3$$
$$\left(\frac{7}{2}\right)y + \left(\frac{17}{2}\right)z = 11$$

Now let us first eliminate y from both the first and third equations. This gives us

$$x - z = 2 \text{ and } z = 5$$

Hence the system becomes:

$$x - z = 2$$
$$y + 3z = 6$$
$$z = 5$$

Evaluation of y and z is trivial and the result is the same as before.

10.7 The LU Factorization

The system to be solved is denoted by $Ax = b$, where A is the coefficient matrix of size $n \times n$. During the process of Gaussian elimination, the linear system was transformed into an upper triangular system denoted as $Ux = g$

$$U = \begin{bmatrix} u_{11} & u_{12} & \cdots & u_{1n} \\ 0 & u_{22} & \cdots & u_{2n} \\ \vdots & & \vdots & \vdots \\ 0 & \cdots & 0 & u_{nn} \end{bmatrix}$$

And $u_{ij} = a_{ij}^{(i)}$. Let's define an auxiliary lower triangular matrix L using the multipliers m_{ij} as follows:

$$U = \begin{bmatrix} 1 & 0 & \cdots & 0 \\ m_{21} & 1 & \cdots & 0 \\ \vdots & & \vdots & \vdots \\ m_{n1} & \cdots & m_{nn-1} & 1 \end{bmatrix}$$

The theorem provides the link between the matrices L and U and the original matrix A.

Theorem 10.1. Let a matrix A be non-singular and L and U be defined as mentioned above. If U is obtained using Gaussian elimination without pivoting, then

$$LU = A$$

which is referred to as the LU factorization of A.

LU factorization provides an alternative viewpoint on Gaussian elimination. Given that $LU = A$, the linear system $Ax = b$ can be expressed as

$$LUx = b\,.$$

And this is equivalent to solving the two systems

$$Lg = b\,, Ux = g \qquad (10.6)$$

The first system is the lower triangular system $g_1 = b_1$

$$m_{21}g_1 + g_2 = b_2$$

$$\vdots$$

$$m_{n1}g_1 + m_{n2}g_2 + \cdots + m_{nn-1}g_{n-1} + g_n = b_n$$

After obtaining g using forward substitution from this system, the upper triangular system $Ux = g$ can be solved using back substitution. Once the factorization $A = LU$ is completed, solving the linear system $Ax = b$ becomes a matter of solving two triangular systems. This significantly reduces the computing cost, especially when solving the problem for a fixed A but with different values of b.

Instead of using the elimination procedures to generate matrices L and U, it is feasible to directly solve for these matrices. Allow us to demonstrate the explicit calculation of L and U for $n = 3$. Express the equation $A = LU$ as

$$\begin{bmatrix} a_{11} & a_{12} & a_{13} \\ a_{21} & a_{22} & a_{23} \\ a_{31} & a_{32} & a_{33} \end{bmatrix} = \begin{bmatrix} 1 & 0 & 0 \\ m_{21} & 1 & 0 \\ m_{31} & m_{32} & 1 \end{bmatrix} \begin{bmatrix} u_{11} & u_{12} & u_{13} \\ 0 & u_{22} & u_{23} \\ 0 & 0 & u_{33} \end{bmatrix} \tag{10.7}$$

The right-hand matrix multiplication gives

$$\begin{gathered} a_{11} = u_{11}, \qquad a_{12} = u_{12}, \qquad a_{13} = u_{13}\,, \\ a_{21} = m_{21}u_{11}, \qquad a_{31} = m_{31}u_{11} \end{gathered} \tag{10.8}$$

Thus, we get first column of L and the first row of U. Next multiply row 2 of L times columns 2 and 3 of U, to get

$$a_{22} = m_{21}u_{12} + u_{22}, \qquad a_{23} = m_{21}u_{13} + u_{23} \tag{10.9}$$

These are solved for u_{12} and u_{12}. After that multiply row 3 of L to obtain

$$m_{31}u_{12} + m_{32}u_{22} = a_{32}\,, m_{31}u_{13} + m_{32}u_{23} + u_{33} = a_{33} \qquad (10.10)$$

These equations yield values for m_{32} and u_{33} , completing the construction of L and U. In this process, we must have $u_{11} \neq 0, u_{22} \neq 0$ in order to solve for L.

It is to be noted that in general the diagonal elements of L need not be 1. The above procedure of LU decomposition is called Doolittle's method.

Example 10.9 Let

$$A = \begin{bmatrix} 1 & 1 & -1 \\ 1 & 2 & -2 \\ -2 & 1 & 1 \end{bmatrix}$$

Using (10.8), we get

$$u_{11} = 1, u_{12} = 1, u_{13} = -1, m_{21} = \frac{a_{21}}{u_{11}} = 1, m_{31} = \frac{a_{31}}{u_{11}} = -2$$

Using (10.9) and (10.10),

$$u_{22} = a_{22} - m_{21}u_{12} = 2 - 1 \times 1 = 1$$
$$u_{23} = a_{23} - m_{21}u_{13} = -2 - 1 \times (-1) = -1$$
$$m_{32} = \frac{(a_{32} - m_{31}u_{12})}{u_{22}} = (1 - (-2) \times 1)/1 = 3$$
$$u_{33} = a_{33} - m_{31}u_{13} - m_{32}u_{23} = 1 - (-2) \times (-1) - 3 \times (-1) = 2$$

Thus,

$$A = \begin{bmatrix} 1 & 0 & 0 \\ 1 & 1 & 0 \\ -2 & 3 & 1 \end{bmatrix} \begin{bmatrix} 1 & 1 & -1 \\ 0 & 1 & -1 \\ 0 & 0 & 2 \end{bmatrix}$$

Taking $b = (1,1,1)$, we now solve the system $Ax = b$ using LU factorization, with the matrix A given above. As discussed above, first we have to solve the lower triangular system

$$\begin{bmatrix} 1 & 0 & 0 \\ 1 & 1 & 0 \\ -2 & 3 & 1 \end{bmatrix} \begin{bmatrix} g_1 \\ g_2 \\ g_3 \end{bmatrix} = \begin{bmatrix} 1 \\ 1 \\ 1 \end{bmatrix}$$

Forward substitution yields $g_1 = 1, g_2 = 0, g_3 = 3$. Keeping the vector $g = (1,0,3)$ as the right-hand side, we now solve the upper triangular system

$$\begin{bmatrix} 1 & 1 & -1 \\ 0 & 1 & -1 \\ 0 & 0 & 2 \end{bmatrix} \begin{bmatrix} x_1 \\ x_2 \\ x_3 \end{bmatrix} = \begin{bmatrix} 1 \\ 0 \\ 3 \end{bmatrix}$$

Backward substitution yields $x_1 = 1\,, x_2 = \frac{3}{2}, x_3 = \frac{3}{2}, \ldots$

Example 10.10 Derive an LU decomposition to solve

$$\begin{gathered} 3x_1 - 0.1x_2 - 0.2x_3 = 7.85 \\ 0.1x_1 + 7x_2 - 0.3x_3 = -19.3 \\ 0.3x_1 - 0.2x_2 + 10x_3 = 71.4 \end{gathered}$$

Solution. The intent of forward substitution is to impose the elimination manipulations, that we had formerly applied to $[A]$, on the right-hand-side vector $\{B\}$.

$$\begin{bmatrix} 3 & -0.1 & -0.2 \\ 0.1 & 7 & -0.31 \\ 0.3 & -0.2 & 10 \end{bmatrix} \begin{bmatrix} x_1 \\ x_2 \\ x_3 \end{bmatrix} = \begin{bmatrix} 7.85 \\ -19.3 \\ 71.4 \end{bmatrix}$$

The forward-elimination phase of conventional Gauss elimination resulted in

$$\begin{bmatrix} 3 & -0.1 & -0.2 \\ 0.1 & 7.00333 & -0.293333 \\ 0 & 0 & 10.0120 \end{bmatrix} \begin{bmatrix} x_1 \\ x_2 \\ x_3 \end{bmatrix} = \begin{bmatrix} 7.85 \\ -19.5617 \\ 70.0843 \end{bmatrix}$$

After the forward-substitution phase is implemented,

$$\begin{bmatrix} 1 & 0 & 0 \\ 0.0333333 & 1 & 0 \\ 0.100000 & -0.0271300 & 1 \end{bmatrix} \begin{bmatrix} d_1 \\ d_2 \\ d_3 \end{bmatrix} = \begin{bmatrix} 7.85 \\ -19.3 \\ 71.4 \end{bmatrix}$$

or multiplying out the left-hand side,

$$d_1 = 7.85$$
$$0.0333333d_1 + d_2 = -19.3$$
$$0.1d_1 - 0.02713d_2 + d_3 = 71.4$$

We can solve the first equation for d_1 ,

$$d_1 = 7.85$$

which can be substituted into the second equation to solve for

$$d_1 = -19.3 - 0.0333333(7.85) = -19.5617$$

Both d_1 and d_2 can be substituted into the third equation to give

$$d_3 = 71.4 - 0.1(7.85) + 0.02713(-19.5617) = 70.0843$$

Thus,

$$\{D\} = \begin{Bmatrix} 7.85 \\ -19.5617 \\ 70.0843 \end{Bmatrix}$$

This result can then be substituted into $[U]\{X\} = \{D\}$, to give

$$\begin{bmatrix} 3 & -0.1 & -0.2 \\ 0.1 & 7.00333 & -0.293333 \\ 0 & 0 & 10.0120 \end{bmatrix} \begin{bmatrix} x_1 \\ x_2 \\ x_3 \end{bmatrix} = \begin{bmatrix} 7.85 \\ -19.5617 \\ 70.0843 \end{bmatrix}$$

which can be solved by back substitution for the final solution,

$$\{X\} = \begin{Bmatrix} 3 \\ -2.5 \\ 7.00003 \end{Bmatrix}$$

Example 10.11 Solve the equations

$$2x + 3y + z = 9$$
$$x + 2y + 3Z = 6$$
$$3x + y + 2z = 8$$

By the method of LU decomposition.

We have $A = \begin{bmatrix} 1 & 0 & 0 \\ 0.5 & 1 & 0 \\ 1.5 & -7 & 1 \end{bmatrix}$ and $B = \begin{bmatrix} 9 \\ 6 \\ 8 \end{bmatrix}$

$L = \begin{bmatrix} 1 & 0 & 0 \\ 0.5 & 1 & 0 \\ 1.5 & -7 & 1 \end{bmatrix}$ and $U = \begin{bmatrix} 2 & 3 & 1 \\ 0 & 0.5 & 2.5 \\ 0 & 0 & 18 \end{bmatrix}$

If $Y = [y_1, y_2, y_3]^T$, then the equation $LY = B$ gives the solution:

$$y_1 = 9\,, y_1 = {}^{3}\!/_{2} \text{ and } y_3 = 5\,.$$

Finally, the matrix equation
$UX = Y$ where $X = [x, y, z]^T$,

Gives the required solution
$x = {}^{35}\!/_{18}\,, y = {}^{29}\!/_{18}$ and $z = {}^{5}\!/_{18}$.

10.8 Matrix Condition and Numerical Stability

This section will examine the impact of issue modifications on the matrix solution process. The primary objective of this endeavor is to get comprehension regarding the challenges associated with solving certain linear system problems. Typically, this is related to a psychological phenomenon called conditioning.

In order to stimulate the discussion, let us examine a straightforward illustration.

Example 10.12 Let A be given by

$$A = \begin{bmatrix} 1.002 & 1 \\ 1 & 0.998 \end{bmatrix}$$

and let b be given by

$$b = \begin{bmatrix} 2.002 \\ 1.998 \end{bmatrix}$$

By doing a straightforward calculation, it can be determined that the solution to the equation $Ax = b$ is $x = (1,1)^T$.

Let us assume that we slightly modify the right-side vector and establish a new definition

$$b_c = \begin{bmatrix} 2.002 \\ 1.998 \end{bmatrix}$$

It should be noted that this represents a change of less than 0.5% in one element of the vector b. Now, a straightforward calculation demonstrates that the solution to the equation $Ax = b_c$ is

$$x_c = \begin{bmatrix} -23.95 \\ 26.00 \end{bmatrix}$$

The most of us probably agree that this represents a significant alteration in the resolution for a comparatively minor modification in the issue. (Furthermore, it is important to highlight that this computation is precise and accurate.) This is a significant challenge in the fields of mathematics and computation. We seek to ensure that minor alterations in a problem will result in proportionally minor changes in the answer. Additionally, we strive to comprehend the reasons behind any exceptions that may arise. We are particularly interested in determining if the significant alteration resulting from a minor disturbance is a result of the problem itself or the computational method used.

Prior to addressing the specific problems in this section, it is necessary to first establish some notation and vocabulary that will enable us to quantify the magnitude of vectors and matrices.

10.8.1 Vector and Matrix Norms

In order to address the measurement of errors in vectors and matrices, it is imperative to introduce the notions of a vector norm and a matrix norm. The vector norm is formally defined as follows.

Definition 10.1 (Vector Norm) A vector norm on $\mathbb{R}^n$ is any mapping $\|\cdot\|$, defined on $\mathbb{R}^n$ with values in $[0, \infty)$, which satisfies the three conditions:

- $\|x\| > 0$ for any vector $x \neq 0$;
- $\|ax\| = |a|\|x\|$ for any scalar a.

- $||x + y|| < ||x|| + ||y||$ for any two vectors x and y.

The two most prevalent examples of vector norms are the infinity norm and the Euclidean 2-norm, which are defined as follows:

$$||x||_\infty = \max_{1 \le i \le n} |x_i|$$

And

$$||x||_2 = \left(\sum_{i=1}^{n} x_i^2 \right)^{\frac{1}{2}}$$

While it is possible to apply the same norm definition to matrix norms, doing so would not yield certain crucial qualities that we want. We use a different kind of matrix norm, referred to as an operator norm.

Definition 10.2 (Vector Norm) Let $||\cdot||$ be a given vector norm defined on $\mathbb{R}^n$. Define the corresponding matrix norm, for matrices $A \in \mathbb{R}^{n \times n}$, by

$$||A|| = \max_{x \neq 0} \frac{||Ax||}{||x||} \tag{10.11}$$

The primary implication of this matrix norm formulation is that the norm of a product is always less than or equal to the product of the norms:

$$||AB|| \le ||A||||B|| \tag{10.12}$$
$$||Ax|| \le ||A||||x|| \tag{10.13}$$

Matrix norms are invariably linked to a specific vector norm, as we have established in this context. Technically, it is not obligatory; it is possible to establish a norm on a matrix without the need for a specific vector norm. Nevertheless, these norms typically fail to meet the requirements outlined in equations (10.12) and (10.13), which are essential for our objectives in this section.

The matrix infinity norm, denoted $||A||_\infty$, can be calculated. The given computation is proven to be equivalent to the following.

$$\|x\|_\infty = \max_{1 \le i \le n} \sum_{j=1}^{n} |a_{ij}|$$

Therefore, $\|A\|_\infty$ is defined as the highest sum of the elements in a row. The matrix 2-norm is more intricate. It can be demonstrated that we have

$$\|A\|_2 = \sqrt{\Lambda(A^T A)},$$

The biggest eigenvalue of matrix B, in terms of absolute value, is denoted as $\Lambda(B)$. Due to the increased complexity in computing using the matrix 2-norm, we will typically choose for the infinity norm.

Example 10.12 Let A be the simple matrix given by

$$A = \begin{bmatrix} 4 & -6 & 2 \\ 0 & 4 & 1 \\ 1 & 2 & 3 \end{bmatrix}$$

Then $\|A\|_\infty = \max\{12,5,6\} = 12$. Furthermore,

$$A^T A = \begin{bmatrix} 17 & -22 & 11 \\ -22 & 56 & -2 \\ 11 & -2 & 14 \end{bmatrix}$$

which has eigenvalues $\lambda = 66.6816, 19.8065, 0.5118$, so that

$$\|A\|_2 = \sqrt{/66.6816} = 8.1659.$$

10.9 Iterative Methods

As previously stated in the chapter's introduction, numerous linear systems are too extensive to be resolved using direct methods using Gaussian elimination. For these systems, iteration methods are typically the most acceptable approach for finding a solution, and they often exceed elimination methods in terms of speed. The primary field where iteration methods are extensively used is in solving linear systems that arise in the numerical solution of partial differential equations. Systems with orders 10^3 to 10^5 are not uncommon, although the majority of the coefficients in the system will be zero.

Besides being large, the linear systems to be solved, $Ax = b$, generally have several other important properties. Typically, they are characterized by sparsity, indicating that just a small percentage of the coefficients have non-zero values. The non-zero coefficients in matrix A typically exhibit a unique pattern and can be generated on-demand using a straightforward method, eliminating the need for storage. One implication of these features is that the store space required for the vectors x and b may be more significant than the storage space required for A. The matrices A frequently possess distinctive characteristics, which will be examined in this section and the following two sections.

Firstly, we establish the definitions and examine two traditional iteration methods. Subsequently, we introduce a comprehensive abstract framework for the analysis of iteration methods.

The unique characteristics of the linear system $Ax = b$ are crucial for establishing an iterative strategy for solving it. The findings of this section represent only the initial stage in the development of a methodology for a certain field of applications.

10.9.1 Jacobi's Method

The Jacobi iterative method is obtained by solving the ith equation in $Ax = b$ for x_i to obtain (provided $a_{ii} \neq 0$)

$$x_i = \sum_{\substack{j=1 \\ j\neq i}}^{n} \left(-\frac{a_{ij}x_j}{a_{ii}}\right) + \frac{b_i}{a_{ii}}, \quad i = 1,2,\dots,n\,.$$

For each $k \geq 1$, generate the components $x_i^{(k)}$ of $x^{(k)}$ from the components of $x^{(k-1)}$ by

$$x_i^{(k)} = \frac{1}{a_{ii}}\left[\sum_{\substack{j=1 \\ j\neq i}}^{n} \left(-a_{ij}x_j{}^{(k-1)}\right) + b_i\right], \qquad i = 1,2,\dots,n. \tag{10.14}$$

Example 10.13 The linear system $Ax = b$ given by

$$10x_1 - x_2 + 2x_3 = 6,$$

$$-x_1 + 11x_2 - x_3 + x_4 = 25,$$
$$2x_1 - x_2 - x_3 + 3x_4 = -11,$$
$$3x_2 - x_3 + 8x_4 = 15,$$

has the unique solution $x = (1,2,-1,1)^t$. Use Jacobi's iterative technique to find approximations $x^{(k)}$ to x starting with $x^{(0)} = (0,0,0,0)^t$ until

$$\frac{\left\|x^{(k)} - x^{(k-1)}\right\|_\infty}{\|x^{(k)}\|_\infty} < 10^{-3}$$

Solution. We first solve equation E_i for x_i, for each $i = 1,2,3,4$, to obtain

$$x_1 = \frac{1}{10}x_2 - \frac{1}{5}x_3 + \frac{3}{5}$$
$$x_2 = \frac{1}{11}x_1 + \frac{1}{11}x_3 - \frac{3}{11}x_4 + \frac{25}{11}$$
$$x_3 = -\frac{1}{5}x_1 + \frac{1}{10}x_3 - \frac{1}{10}x_4 - \frac{11}{10}$$
$$x_4 = -\frac{3}{8}x_2 + \frac{1}{8}x_3 + \frac{15}{8}$$

From the initial approximation $x^{(0)} = (0,0,0,0)^t$ we have $x^{(1)}$ given by

$$x_1^{(1)} = \frac{1}{10}x_2^{(0)} - \frac{1}{5}x_3^{(0)} + \frac{3}{5} = 0.6000$$
$$x_2^{(1)} = \frac{1}{11}x_1^{(0)} - \frac{1}{5}x_3^{(0)} - \frac{3}{11}x_4^{(0)} + \frac{25}{11} = 2.2727$$
$$x_3^{(1)} = -\frac{1}{5}x_1^{(0)} + \frac{1}{10}x_2^{(0)} + \frac{1}{10}x_4^{(0)} - \frac{11}{10} = -1.1000$$
$$x_4^{(1)} = -\frac{3}{8}x_2^{(0)} + \frac{1}{8}x_3^{(0)} + \frac{15}{8} = 1.8750$$

Additional iterates, $x^{(k)} = \left(x_1^{(k)}, x_2^{(k)}, x_3^{(k)}, x_4^{(k)}\right)$ are generated in a similar manner and are presented in Table 10.2.

Table 10.2 Additional iterates for Example 10.13

1	2	3	4	5	6	7	8	9	10
0.6000	1.0473	0.9326	1.0152	0.9890	1.0032	0.9981	1.0006	0.9997	1.0001
2.2727	1.7159	2.053	1.9537	2.0114	1.9922	2.0023	1.9987	2.0004	1.9998
−1.100	−0.805	−1.049	−0.968	−1.010	−0.994	−1.002	−0.999	−1.000	−0.999
1.8750	0.8852	1.1309	0.9739	1.0214	0.9944	1.0036	0.9989	1.0006	0.9998

We stopped after ten iterations because

$$\frac{\left\|x^{(10)} - x^{(9)}\right\|_\infty}{\|x^{(10)}\|_\infty} = \frac{8 \times 10^{-4}}{1.9998} < 10^{-3}$$

In fact, $\left\|x^{(10)} - x\right\|_\infty = 0.0002$.

In general, iterative techniques for solving linear systems involve a process that converts the system $Ax = b$ into an equivalent system of the form $x = Tx + c$ for some fixed matrix T and vector c. After the initial vector $x^{(0)}$ is selected, the sequence of approximate solution vectors is generated by computing

$$x^{(k)} = Tx^{(k-1)} + c,$$

for each $k = 1, 2, 3, \ldots$. This should be reminiscent of the fixed-point iteration. The Jacobi method can be written in the form $x^{(k)} = Tx^{(k-1)} + c$, by splitting A into its diagonal and off-diagonal parts. To see this, let D be the diagonal matrix whose diagonal entries are those of A, $-L$ be the strictly lower-triangular part of A, and $-U$ be the strictly upper-triangular part of A. With this notation,

$$A = \begin{bmatrix} a_{11} & a_{12} & \cdots & a_{1n} \\ a_{21} & a_{22} & \cdots & a_{2n} \\ \vdots & \vdots & & \vdots \\ a_{n1} & a_{n2} & \cdots & a_{nn} \end{bmatrix}$$

is split into

$$A = \begin{bmatrix} a_{11} & 0 & \cdots & 0 \\ 0 & a_{22} & \ddots & \vdots \\ \vdots & \ddots & \ddots & 0 \\ 0 & \cdots & 0 & a_{nn} \end{bmatrix} - \begin{bmatrix} 0 & \cdots & \cdots & a_{1n} \\ -a_{21} & \ddots & \ddots & \vdots \\ \vdots & \ddots & \ddots & \vdots \\ -a_{n1} & \cdots & -a_{n,n-1} & 0 \end{bmatrix} - \begin{bmatrix} 0 & -a_{12} & \cdots & -a_{1n} \\ \vdots & \ddots & \ddots & \vdots \\ \vdots & \ddots & \ddots & -a_{n-1,n} \\ 0 & \cdots & \cdots & 0 \end{bmatrix} = D - L - U$$

The equation $Ax = b$, or $(D - L - U)x = b$, is then transformed into

$$Dx = (L + U)x + b$$

and, if D^{-1} exists, that is, if $a_{ii} \neq 0$ for each i, then

$$x = D^{-1}(L + U)x + D^{-1}b$$

This results in the matrix form of the Jacobi iterative technique:

$$x^{(k)} = D^{-1}(L + U)x^{(k-1)} + D^{-1}b\,, \quad k = 1,2, \ldots \tag{10.15}$$

Introducing the notation $T_j = D^{-1}(L + U)$ and $c_j = D^{-1}b$ gives the Jacobi technique the form

$$x^{(k)} = T_j x^{(k-1)} + c_j \tag{10.16}$$

In practice, Eq. (10.14) is used in computation and Eq. (10.16) for theoretical purposes.

Example 10.14 Express the Jacobi iteration method for the linear system $Ax = b$ given by

$$\begin{aligned} 10x_1 - x_2 + 2x_3 &= 6, \\ -x_1 + 11x_2 - x_3 + x_4 &= 25, \\ 2x_1 - x_2 - x_3 + 3x_4 &= -11, \\ 3x_2 - x_3 + 8x_4 &= 15, \end{aligned}$$

in the form $x^{(k)} = T_j x^{(k-1)} + c_j$.

Solution. We saw in Example 10.13 that the Jacobi method for this system has the form

$$x_1 = \frac{1}{10}x_2 - \frac{1}{5}x_3 + \frac{3}{5}$$

$$x_2 = \frac{1}{11}x_1 + \frac{1}{11}x_3 - \frac{3}{11}x_4 + \frac{25}{11}$$

$$x_3 = -\frac{1}{5}x_1 + \frac{1}{10}x_3 - \frac{1}{10}x_4 - \frac{11}{10}$$

$$x_4 = -\frac{3}{8}x_2 + \frac{1}{8}x_3 + \frac{15}{8}$$

Hence, we have

$$T = \begin{bmatrix} 0 & 1/10 & -1/5 & 0 \\ 1/11 & 0 & 1/11 & -3/11 \\ -1/5 & 1/10 & 0 & 1/10 \\ 0 & -3/8 & 1/8 & 0 \end{bmatrix} \text{ and } c = \begin{bmatrix} 3/5 \\ 25/11 \\ -11/10 \\ 15/8 \end{bmatrix}.$$

10.9.2 The Gauss-Seidel Method

Iterative or approximation approaches offer an alternative to the elimination methods outlined thus far. These methods involved making an initial estimation and then using a systematic strategy to acquire a more accurate estimate of the root. Given that the current section of the book addresses a comparable issue - the acquisition of values that fulfil a set of equations concurrently - it is reasonable to believe that these approximate methods could be advantageous in this particular situation.

The Gauss-Seidel method is the predominant iterative method. Let us consider a given set of n equations:

$$[A]\{X\} = \{B\}$$

Let's assume a set of equations with a $3x3$ matrix for the purpose of conciseness. If all the diagonal elements are non-zero, the first equation can be solved for x_1, the second for x_2, and the third for x_3 to obtain the solution

$$x_1 = \frac{b_1 - a_{11}x_2 - a_{13}x_3}{a_{11}} \tag{10.17a}$$

$$x_2 = \frac{b_2 - a_{21}x_1 - a_{23}x_3}{a_{22}} \tag{10.17b}$$

$$x_3 = \frac{b_3 - a_{31}x_1 - a_{33}x_2}{a_{33}} \tag{10.17c}$$

Now, we can start the solution process by choosing guesses for the x's. A simple way to obtain initial guesses is to assume that they are all zero. These zeros can be substituted into Eq. (10.17a), which can be used to calculate a new value for $x_1 = b_1/a_{11}$. Then, we substitute this new value of x_1 along with the previous guess of zero for x_3 into Eq. (10.17b) to compute a new value for x_2. The process is repeated for Eq. (10.17c) to calculate a new estimate for x_3. Then we return to the first equation and repeat the entire procedure until our solution converges closely enough to the true values. Convergence can be checked using the criterion.

$$|\varepsilon_{a,i}| = \left|\frac{x_i^j - x_i^{j-1}}{x_i^j}\right| 100\% < \varepsilon_s \tag{10.18}$$

for all i, where j and $j-1$ are the present and previous iterations.

Example 10.15 Use the Gauss-Seidel method to obtain the solution of the system

$$3x_1 - 0.1x_2 - 0.2x_3 = 7.85$$
$$0.1x_1 + 7x_2 - 0.3x_3 = -19.3$$
$$0.3x_1 - 0.2x_2 + 10x_3 = 71.4$$

Solution. First, solve each of the equations for its unknown on the diagonal.

$$x_1 = \frac{7.85 + 0.1x_2 + 0.2x_3}{3} \qquad \text{(i)}$$

$$x_2 = \frac{-19.3 - 0.1x_1 + 0.3x_3}{7} \qquad \text{(ii)}$$

$$x_3 = \frac{71.4 - 0.3x_1 + 0.2x_2}{10} \qquad \text{(iii)}$$

By assuming that x_2 and x_3 are zero, Eq. (i) can be used to compute

$$x_1 = \frac{7.85 + 0 + 0}{3} = 2.616667$$

This value, along with the assumed value of $x_3 = 0$, can be substituted into Eq. (ii) to calculate

$$x_2 = \frac{-19.3 - 0.1(2.616667) + 0}{7} = -2.794524$$

The first iteration is completed by substituting the calculated values for x_1 and x_2 into Eq. (iii) to yield

$$x_3 = \frac{71.4 - 0.3(2.616667) + 0.2(-2.794524)}{10} = 7.005610$$

For the second iteration, the same process is repeated to compute

$$x_1 = \frac{7.85 + 0.1(-2.794524) + 0.2(7.005610)}{3} = 2.990557 \quad |\varepsilon_t| = 0.31\%$$

$$x_2 = \frac{-19.3 - 0.1(2.990557) + 0.3(7.005610)}{7} = -2.499625 \quad |\varepsilon_t| = 0.015\%$$

$$x_3 = \frac{71.4 - 0.3(2.990557) + 0.2(-2.499625)}{10} = 7.000291 \quad |\varepsilon_t| = 0.0042\%$$

The method is approaching the accurate solution. Further iterations may be employed to enhance the responses. However, in an actual situation, the true answer would not be known in advance. Therefore, Equation (10.18) offers a method to approximate the error.

As an example, for x_1,

$$|\varepsilon_{a,1}| = \left|\frac{2.990557 - 2.616667}{2.990557}\right| 100\% = 12.5\%$$

The error estimates for x_2 and x_3 are $|\varepsilon_{a,2}| = 11.8\%$ and $|\varepsilon_{a,3}| = 0.076\%$, respectively. It is important to note that, similar to finding roots of a single equation, equations like Eq. (10.18) typically offer a cautious evaluation of convergence. Therefore, when these conditions are fulfilled, they guarantee that the outcome is understood to the level of precision specified by ε_s.

10.10 Exercises

1. Write the following set of equations in matrix form:

$$8 = 6x_3 + 2x_2$$
$$2 - x_1 = x_3$$
$$5x_2 + 8x_1 = 13$$

2. Solve by Gauss elimination

$$10x_1 + 2x_2 - x_3 = 27$$
$$-3x_1 - 6x_2 + 2x_3 = -61.5$$
$$x_1 + x_2 + 5x_3 = -21.5$$

3. Use Gauss elimination to solve:

$$8x_1 + 2x_2 - 2x_3 = -2$$
$$10x_1 + 2x_2 + 4x_3 = 4$$
$$12x_1 + 2x_2 + 2x_3 = 6$$

Employ partial pivoting and check your answers by substituting them into the original equations.

4. Use Gauss elimination with partial pivoting to solve for the x's.

$$-3x_2 + 7x_3 = 2$$
$$x_1 + 2x_2 - x_3 = 3$$
$$5x_1 - 2x_2 = 2$$

5. Given the equations

$$2x_1 - 6x_2 - x_3 = -38$$
$$-3x_1 - x_2 + 7x_3 = -34$$
$$-8x_1 + x_2 - 2x_3 = -20$$

(a) Solve by Gauss elimination with partial pivoting. Show all steps of the computation.
(b) Substitute your results into the original equations to check your answers.

6. Use Gauss-Jordan elimination to solve:

$$2x_1 + x_2 - x_3 = 1$$
$$5x_1 + 2x_2 + 2x_3 = -4$$
$$3x_1 + x_2 + 2x_3 = 5$$

Do not employ pivoting. Check your answers by substituting them into the original equations.

7. Solve

$$x_1 + x_2 - x_3 = -3$$
$$6x_1 + 2x_2 + 2x_3 = 2$$
$$-3x_1 + 4x_2 + x_3 = 1$$

(a) With Gauss elimination with partial pivoting.
(b) Gauss-Jordan without partial pivoting.

8. Solve the following system of equations by LU decomposition with partial pivoting

$$2x_1 - 6x_2 - x_3 = -38$$
$$-3x_1 - x_2 + 7x_3 = -34$$
$$-8x_1 + x_2 - 2x_3 = -20$$

9. Solve the following system of equations by LU decomposition without pivoting

$$8x_1 + 4x_2 - x_3 = 11$$
$$-2x_1 + 5x_2 + x_3 = 4$$
$$2x_1 - x_2 + 6x_3 = 7$$

10. Determine the total flops as a function of the number of equations n for the (a) decomposition, (b) forward-substitution, and (c) back-substitution phases of the LU decomposition version of Gauss elimination.

11. Use LU decomposition to determine the matrix inverse for the following system. Do not use a pivoting strategy, and check your results by verifying that $[A][A]^{-1} = [I]$.

$$10x_1 + 2x_2 - x_3 = 27$$
$$-3x_1 - 6x_2 + 2x_3 = -61.5$$
$$x_1 + x_2 + 5x_3 = -21.5$$

12. Solve the following set of equations with LU decomposition:

$$3x_1 - 2x_2 + x_3 = -10$$
$$2x_1 + 6x_2 - 4x_3 = 44$$
$$-x_1 - 2x_2 + 5x_3 = -26$$

13. Use the Gaussian elimination method with rounding up to 5 digits after decimal point to solve the system

$$0.52x_1 - x_2 = -0.5$$
$$1.02x_1 - 2x_2 = -18.8$$

14. For any $n \times n$ matrices A and B, and $x \in \mathbb{R}^n$, show that

(a) $\|Ax\| \leq \|A\|\|x\|$

(b) $\|AB\| \leq \|A\|\|B\|$

where the matrix norm is the induced norm obtained from the corresponding vector norm.

15. Solve the system

$$5x_1 + 7x_2 = b_1$$
$$7x_1 + 10x_2 = b_2$$

using Gaussian elimination method to obtain the solution x_1 when $b_T = (b_1, b_2) = (0.7, 1)$. Also solve the above system with $b_A = (b_1, b_2) = (0.69, 1.01)$ using the same method to obtain the solution x_2.

16. Find the $n \times n$ matrix B and the n-dimensional vector c such that the Gauss-Seidal method can be written in the form

$$x^{(k+1)} = Bx^{(k)} + c\,, \qquad k = 1,2,\ldots$$

17. Show that the Gauss-Seidal method converges if the coefficient matrix is diagonally dominant.

18. Use Jacobi method to perform 3 iterations with $x^{(0)} = (0,0,0)$ to get $x^{(1)}, x^{(2)}$ and $x^{(3)}$ for the system

$$-x_1 + 5x_2 - 2x_3 = 3$$
$$x_1 + x_2 - 4x_3 = -9$$
$$4x_1 - x_2 + 2x_3 = 8$$

19. Solve the system

$$6x + y + z = 20$$
$$x + 4y - z = 6$$
$$x - y + 5z = 7$$

Using both Jacobi and Gauss-Seidel methods.

20. Express the following systems of equations I the matrix form:

(a) $3x + 2y + 4z = 7$

$$2x + y + z = 7$$
$$x + 3y + 5z = 2$$

(b) $2x - y + z = 20$
$$x + 4y - z = 6$$
$$-x + 3y + 5z = 7$$

21. The linear system

$$2x_1 - x_2 + x_3 = -1$$
$$2x_1 + 2x_2 + 2x_3 = 4$$
$$-x_1 - x_2 + 2x_3 = -5$$

has the solution $(1, 2, -1)^t$.

(a) Show that the Jacobi method with $x^{(0)} = 0$ fails to give a good approximation after 25 iterations.

(b) Use the Gauss-Seidel method with $x^{(0)} = 0$ to approximate the solution to the linear system to within 10^{-5} in the l_∞ norm.

CHAPTER 11

The Finite Element Method

11.1 Introduction

In previous chapters, we studied finite difference methods for solving boundary-value problems defined by ordinary and partial differential equations. Here, we will discuss a different set of techniques called finite element methods that are used to solve these types of problems. Finite element analysis (FEA) involves using calculations made with the finite element method (FEM) to predict the behaviour of an object. Analytical approaches are generally ineffective in solving Partial Differential Equations for most geometries and problems. Alternatively, we can use discretisation methods to approximate these equations, which can then be solved using numerical methods. Thus, the solutions obtained are an approximation of the actual solution to the PDEs. The Finite Element Method (FEM) is an approximation method that divides a complex space or domain into small, countable, and finite pieces. These pieces, known as finite elements, can be described using simple equations.

FEM uses math to break complex systems into smaller, simpler pieces, or "elements." It then applies differential equations to each element individually, using the power of computers to divide, then conquer engineering problems. In this chapter, we will discuss the finite element method, which is a significant numerical application of the Rayleigh-Ritz and Galerkin methods. The mathematical software is widely used in engineering and applied science to solve practical problems. The finite element method involves subdividing the integration domain into smaller regions called elements. Within each element, the continuous function is approximated using a piecewise polynomial. In order to achieve a more accurate approximation, it is not necessary to utilise higher-order polynomials. Instead, one can simply employ a finer subdivision, such as increasing the number of elements.

Various types of elements are commonly used in practice, with the specific type chosen based on the geometric shape of the region being studied. Triangles, rectangles, and quadrilaterals are commonly

used in two-dimensional problems. Tetrahedra, hexahedra, and parallelepiped elements are commonly employed for three-dimensional problems. In this chapter, we aim to provide an introduction to the finite element method. Therefore, we will focus on its application to solving simple one-dimensional problems. Fig. 11.1 displays examples of common finite elements.

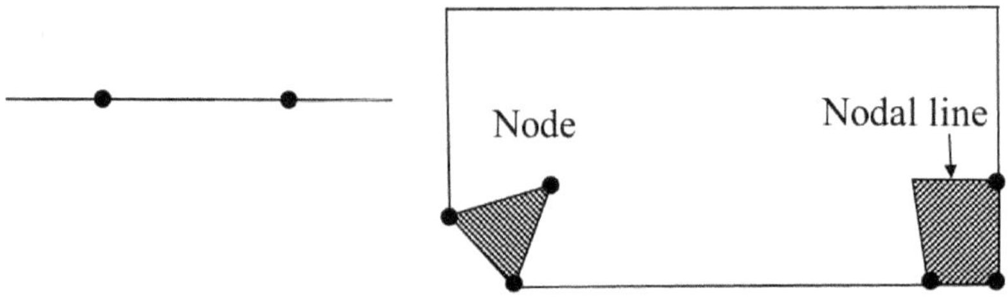

(a) Line element (b) Triangular and Quadrilateral elements

Figure 0.1 Typical finite elements

11.1.1 Functionals

In order to understand the Rayleigh-Ritz method, it is necessary to have a learn of the concept of a functional. This will be further explored in the upcoming section. This concept is commonly found in the study of variational principles, which are prevalent in various physical and other problems. In mathematical terms, a variational principle involves finding the extreme value of an integral for a specific function, denoted as $f(x, y, y')$. The integrand is a function of the coordinates and their derivatives, and the integration is carried out over a specific region.

$$I(y) = \int_{a}^{b} f(x, y, y')\, dx \tag{11.1}$$

Take, for instance, the integral given by $b = f(x, y, y')dx$, a where y(x) fulfils the boundary conditions $y(a) = y(b) = 0$.

The function is integrated over the one-dimensional domain x. It is commonly understood that a functional is a mathematical concept that takes a function y and produces a real number, specifically the

value of the definite integral in Eq. (11.1). According to the calculus of variations, it is established that for $I(y)$ to possess an extremum, $y(x)$ must adhere to the Euler-Lagrange differential equation:

$$\frac{\partial f}{\partial y} - \frac{d}{dx}\left(\frac{\partial f}{\partial y'}\right) = 0 \tag{11.2}$$

In a similar manner, when considering functionals of this particular form

$$I(y) = \int_a^b f(x, y, y', y'')\, dx \tag{11.3}$$

Functionals of the form (11.2) can be analysed using the Euler-Lagrange equation, which is expressed as (11.4). Similarly, other functionals, such as those mentioned in (11.3), can also be examined using this equation.

$$\frac{\partial f}{\partial y} - \frac{d}{dx}\left(\frac{\partial f}{\partial y'}\right) + \frac{\partial f}{\partial y} - \frac{d^2}{dx^2}\left(\frac{\partial f}{\partial y''}\right) = 0 \tag{11.4}$$

The Euler-Lagrange equation (11.2) has several solutions and the one which satisfies the given boundary conditions is selected. Thus, one determines the functional so that it takes on an extremum value from a set of permissible functions. This is the central problem of a variational principle. An important point here is that an extremum may not exist. In other words, a variational principle may exist, but an extremum may not exist. Furthermore, not all differential equations have a variational principle. These difficulties are serious and therefore impose limitations on the application of the variational principle to the solution of engineering problems. Many problems arising in physics and engineering are modelled by boundary- value problems and initial boundary-value problems. Frequently, these equations are equivalent to the problem of the minimization of a functional which can be interpreted in terms of the total energy of the given system. In any physical situation, therefore, the functional is obtained from a consideration of the total energy explicitly. Mathematically, however, it would be useful to be able to determine the functional from the governing differential equation itself. This is illustrated below with an example.

Example 11.1 Find the functional for the boundary value problem defined by

$$\frac{d^2}{dx^2} = f(x) \qquad \text{(i)}$$

And $$y(a) = y(b) = 0 \qquad \text{(ii)}$$

We have

$$\delta \int_a^b f\, y\, dx = \int_a^b f\delta\, y\, dx$$

$$= \int_a^b \frac{d^2y}{dx^2}\,\delta y\, dx \text{ , since } f(x) = \frac{d^2y}{dx^2}$$

$$= \left[\frac{dy}{dx}\delta y\right]_a^b - \int_a^b \frac{dy}{dx}\frac{d}{dx}(\delta y)dx \text{ ,}$$

Om integrating by parts

$$= -\int_a^b \frac{dy}{dx}\frac{d}{dx}(\delta y)dx \text{ , since } \delta y(a) = \delta y(b) = 0$$

$$= -\int_a^b \frac{dy}{dx}\delta\left(\frac{dy}{dx}\right)dx \text{ , since } \frac{d}{dx}(\delta y) = \delta\left(\frac{d}{dx}\right)$$

$$= -\int_a^b \frac{1}{2}\delta\left(\frac{dy}{dx}\right)^2 dx$$

$$= -\delta\int_a^b \frac{1}{2}\left(\frac{dy}{dx}\right)^2 dx$$

Hence

$$\delta\int_a^b \left[fy + \frac{1}{2}\left(\frac{dy}{dx}\right)^2\right]dx = 0$$

It follows that a unique solution of the problem (i) to (ii) exists at a minimum value of the integral defined by

$$I(v) = \int_a^b \left[fv + \frac{1}{2}\left(\frac{dy}{dx}\right)^2 \right] dx \qquad \text{(iii)}$$

Consider the function $v(x)$ that fulfils the necessary boundary conditions, specifically $v(a) = v(b) = 0$. Compute the product of the differential equation expressed in the given form $-y'' + f(x) = 0$ By v and integrate with respect to x. We then get

$$\begin{aligned} 0 &= -\int_a^b vy''\,dx + \int_a^b vf\,dx \\ &= [-vy']_a^b + \int_a^b v'y'\,dx + \int_a^b vf\,dx \\ &= \int_a^b (v'y' + vf)\,dx \end{aligned}$$

Lastly, replace y with v in the previous equation and then multiply the bilinear terms by one-half. Next, we acquire the necessary functional

$$I(v) = \int_a^b \left[\frac{1}{2}v'^2 + vf\right] dx$$

Which is equivalent to the equation. Thirdly Hence, the integral in equation (iii) symbolises the necessary functional of the issue, as per its specification. Similarly, functionals of additional boundary-value and beginning conditions The book does not thoroughly cover the determination of functionals relating to boundary-value problems. Below, we provide a list of known boundary-value problems together with their corresponding functionals. This information will be helpful in comprehending the difficulties presented in this chapter.

(i) $$\frac{d^2y}{dx^2} = f(x), \quad y(a) = y(b) = 0 \qquad (11.5)$$

$$I(v) = \int_a^b v(2f - v'')\,dx \tag{11.6}$$

(ii) $$\frac{d^2y}{dx^2} + ky = x^2, 0 < x < 1;\; y(0) = 0,$$

$$\left(\frac{dy}{dx}\right)_{x=1} = 1 \tag{11.7}$$

$$I(v) = \frac{1}{2}\int_a^b \left[\left(\frac{dy}{dx}\right)^2 - kv^2 + 2vx^2\right] dx - v(1) \tag{11.8}$$

(iii) $$x^2y'' + 2xy' = f(x),$$

$$y(a) = y(b) = 0 \tag{11.9}$$

$$I(v) = \int_a^b v\left[2f - \frac{d}{dx}(x^2y')\right] dx \tag{11.10}$$

(iv) $\nabla^2 u = 0, u = 0$ on the boundary C of R.

(11.11)

$$I(v) = \iint_R \frac{1}{2}\left[\left(\frac{\partial u}{\partial x}\right)^2 + \left(\frac{\partial u}{\partial y}\right)^2\right] dxdy \tag{11.12}$$

(v) $\nabla^2 u = -f, u = 0$ on the boundary C of R. (11.13)

$$I(v) = \iint_R \left\{\frac{1}{2}\left[\left(\frac{\partial u}{\partial x}\right)^2 + \left(\frac{\partial u}{\partial y}\right)^2\right] - uf\right\} dxdy \tag{11.14}$$

(vi) $$\left.\begin{aligned} EI\frac{d^4y}{dx^4} + ky = f(x), 0 < x < l \\ y = 0 = \frac{d^2y}{dx^2} \text{ at } x = 0, l \end{aligned}\right\} \tag{11.15}$$

$$I(v) = \frac{1}{2}\int_0^l \left[EI\left(\frac{d^2v}{dx^2}\right)^2 + kv^2 - 2vf\right] dx \tag{11.16}$$

11.2 Weighted Residual Method

This section focuses on two methods of approximation: The Rayleigh-Ritz and Galerkin methods. As previously stated, the former approach relies on the presence of a functional that is subsequently minimised. Galerkin introduced the second technique as a means for minimising errors. It is part of a broader category of techniques known as weighted residual approaches. The Weighted Residual Method (WRM) is a category of methods employed to derive an estimated solution for differential equations with the form

$$\mathcal{L}(\phi) + f = 0 \text{ in } D \tag{11.17}$$

Within the framework of WRM, we engage directly with the differential equation associated with the problem, without depending on any variational principle. It is equally appropriate for both linear and nonlinear differential equations. The weighted residual approach consists of two primary steps. Initially, we make an assumption of an approximate solution by considering the overall pattern of the dependent variable. The solution is chosen to be approximate that it fulfils the boundary criteria for ϕ. The presumed solution is subsequently replaced in the differential equation. Due to its approximate nature, the supposed solution fails to meet the requirements of the differential equation, resulting in an inaccuracy known as a residual. The residual is thereafter eliminated in a comprehensive manner across the full solution area. This process yields a set of algebraic equations. The next step involves solving the system of equations that arise from the previous step, while considering the specified boundary condition. This will provide the desired approximate solution.

11.3 Rayleigh–Ritz Method

The Rayleigh-Ritz method is a direct method used to minimise a given functional. The solution to the variational problem is obtained without the need to solve the associated Euler-Lagrange Equation, making it a straightforward approach. It is worth mentioning that, in the case of most physical problems, the functional obtained from the variational principle is not straightforward, making it challenging to obtain a solution using the Euler-Lagrange equation. The Rayleigh–Ritz method is an approximate technique that directly minimises the given functional, without needing to rely on the associated Euler-Lagrange equation.

Consider the following functional examples that illustrate the method.

$$J[\phi] = \int_S F(x, y, \phi, \phi_x, \phi_y) dS \tag{11.18}$$

Our goal is to minimise this integral. In the Rayleigh-Ritz method, a set of linearly independent basis functions called un are chosen to construct an approximate solution to equation (11.18), while ensuring that certain boundary conditions are satisfied. The solution can be expressed as a finite series

$$\tilde{\phi} = u_0 + \sum_{n=1}^{N} a_n u_n \tag{11.19}$$

The point where u_0 intersects the nonhomogeneous boundary conditions, if applicable, and un fulfils the homogeneous boundary conditions. The coefficients a need to be determined and $\tilde{\phi}$ is an approximate solution to the exact solution ϕ. When the approximate solution is substituted into equation (11.18), a function with N coefficients $a_1, a_2, \cdots a_N$ is obtained.

$$J(\tilde{\phi}) = J(a_1, a_2, \cdots a_N)$$

The function reaches its minimum when the partial derivatives with respect to each coefficient are zero. As such,

$$\frac{\partial J}{\partial a_1} = 0, \quad \frac{\partial J}{\partial a_2} = 0, \quad \cdots \quad \frac{\partial J}{\partial a_N} = 0$$
$$\frac{\partial J}{\partial a_n} = 0, n = 1,2, \ldots N \tag{11.20}$$

Therefore, we can derive a system of N linear algebraic equations that can be solved to obtain a solution. The given values are then inserted into the approximate solution (11.19). If $\tilde{\phi}$ approaches ϕ as N

approaches infinity, the procedure is considered to converge to the exact solution.

The selection of basic functions is made to meet the specified boundary conditions of the problem.

In order to meet the inhomogeneous boundary conditions, u0 is carefully selected, while $u_n (n = 1,2,\cdots N)$ are chosen to fulfil the homogeneous boundary conditions. It is worth mentioning that when all the prescribed boundary conditions are homogeneous (Dirichlet conditions), $u_0 = 0$. There are two significant limitations associated with the Rayleigh-Ritz method. In certain cases, such as when dealing with nonself-adjoint equations involving odd order derivatives, the variational principle described in equation (11.18) may not be applicable. Furthermore, it can be quite challenging, if not exceedingly difficult, to locate the appropriate functions u0 that satisfy the global boundary conditions for domains with intricate geometries.

Example 11.2 We consider the two-point boundary value problem defined by

$$y'' + x = 0\,, 0 < x < 1, \qquad y(0) = y(1) = 0 \tag{i}$$

Solution. From Eq. (11.6), we obtain,

$$I(v) = \int_0^1 v(-2x - v'')dx = -\int_0^1 v(2x + v'')dx \tag{ii}$$

Let

$$v(x) = \textstyle\sum_{i=1}^{n} \alpha_i \phi_i(x) \tag{iii}$$

Where

$$\phi_i(0) = \phi_i(1) = 0 \text{ for all } i. \tag{iv}$$

Substituting (iii) in (ii), we obtain

$$I(v) = -\int_0^1 \left[\sum_{i=1}^{n} \alpha_i \phi_i(x)\right]\left[2x + \sum_{j=1}^{n} \alpha_j \phi_j''(x)\right] dx \tag{v}$$

For convenience, we set

$$P_i = \textstyle\int_0^1 x\phi_i(x)dx \tag{vi}$$

And

$$q_{ij} = \int_{0}^{1} \phi_i(x)\phi_j''(x)dx$$

$$= \left[\phi_i(x)\phi_j'(x)\right]_0^1 - \int_0^1 \phi'_i(x)\phi_j'(x)dx \text{, on integrating by parts}$$

$$= -\int_0^1 \phi'_i(x)\phi_j'(x) \tag{vii}$$

Using boundary conditions (iv).

Then Eq. (v) becomes

$$I(v) = -2\sum_{i=1}^{n} \alpha_i p_i - \sum_{i=1}^{n}\sum_{j=1}^{n} \alpha_i\alpha_j q_{ij}$$

Hence ${\partial I}/{\partial \alpha_i} = 0$ gives

$$2p_i + 2\sum_{j=1}^{n} \alpha_j q_{ij} = 0\,, \quad (i = 1,2, \dots, n). \tag{viii}$$

Since, $n = 2$ and we therefore choose $\phi_1(x) = x(x-1)$ and $\phi_2(x) = x^2(1-x)$, so that the boundary conditions (iv) are satisfied. Now, from (vi),

$$p_1 = \int_{0}^{1} x^2(1-x)dx = \frac{1}{12}$$

And

$$p_2 = \int_{0}^{1} x^3(1-x)dx = \frac{1}{20}$$

Also, $\phi_1'(x) = 1 - 2x$ and $\phi_2'(x) = 2x - 3x^2$. Eq. (vii) gives

$$q_{11} = -\int_0^1 (1-2x^2)dx = -\frac{1}{3}$$

$$q_{11} = -\int_0^1 (1-2x)(2x-3x^2)dx = -\frac{1}{6} = q_{21}$$

$$q_{22} = -\int_0^1 (2x-3x^2)^2 dx = -\frac{2}{15}.$$

Eq. (viii) now gives

$4\alpha_1 + 2\alpha_2 = 1$ and $10\alpha_1 + 8\alpha_2 = 3$,

Whose solution is $\alpha_1 = \alpha_2 = \frac{1}{6}$. Hence

$$v(x) = \frac{1}{6}x(1-x) + \frac{1}{6}x^2(1-x) = \frac{1}{6}x(1-x^2).$$

It can be verified that this is the exact solution of the problem (i).

Example 11.3 Find Solution of boundary-value problem defined by

$$y'' - x = 0 \tag{i}$$

And

$$y(0) = 0,\ y'(1) = -\frac{1}{2} \tag{ii}$$

By the Rayleigh-Ritz method.

Solution. In this case, one of the boundary conditions is essential and the other natural. Also, the exact solution of the problem is given by

$$y(x) = \frac{x^3}{6} - x\,. \tag{iii}$$

The functional for this problem is given by

$$I(v) = \int_0^1 \left(v'^2 + 2vx\right)dx + v(1)\,. \tag{iv}$$

Let

$$v(x) = \alpha_1 x + \alpha_2 x^2 \tag{v}$$

be an approximate solution so that $v(x)$ satisfies the essential boundary condition, viz., $v(0) = 0$. Then $v'(x) = \alpha_1 + 2\alpha_2 x$ and Eq. (iv) gives

$$I(v) = \int_0^1 [(\alpha_1 + 2\alpha_2 x)^2 + 2x(\alpha_1 x + \alpha_2 x^2)]\,dx + \alpha_1 + \alpha_2 \tag{vi}$$

Hence,

$$\left.\begin{aligned}\frac{\partial I}{\partial \alpha_1} = 0 = \int_0^1 [2(\alpha_1 + 2\alpha_2 x) + 2x^2]\, dx + 1 \\ \frac{\partial I}{\partial \alpha_2} = 0 = \int_0^1 [2(\alpha_1 + 2\alpha_2 x)2x + 2x^3]\, dx + 1.\end{aligned}\right\} \qquad \text{(vii)}$$

Simplification gives the two equations

$$\alpha_1 + \alpha_2 = -{}^5/_6 \text{ and } \alpha_1 + {}^4/_3\,\alpha_2 = -{}^3/_4, \qquad \text{(viii)}$$

Whose solution is $\alpha_1 = -{}^{13}/_{12}$ and $\alpha_2 = {}^1/_4$.

The approximate solution is given by

$$y^{(1)} = -\frac{13}{12}x + \frac{1}{4}x^2 .$$

The student should compare this with the exact solution.

11.4 Galerkin Method

In the Galerkin version of the weighted residual technique, the weight functions are selected to be the same as the trial functions. This is the conventional approach we typically employ to formulate finite element equations for field problems. In the Galerkin method, we establish

$$w_i = N_i\,, \qquad (i = 1,2, \dots, n). \qquad (11.21)$$

The undetermined coefficients in the approximate solution are obtained by equating the integral over D of the weighted residual to zero. This approach will yield results for a one-dimensional problem inside the interval $[a, b]$.

$$\int_a^b w_i R(x)dx = \int_a^b N_i R(x)dx = 0\,, \qquad (i = 1,2, \dots, n)$$

$$(11.22)$$

The Galerkin approach results in a linear system of equations that can be expressed in the following form:

$$\begin{bmatrix} \int_a^b N_1\mathcal{L}(N_1)\,dx & \int_a^b N_1\mathcal{L}(N_2)\,dx & \cdots & \int_a^b N_1\mathcal{L}(N_n)\,dx \\ \cdots & \cdots & \cdots & \cdots \\ \cdots & \cdots & \cdots & \cdots \\ \int_a^b N_n\mathcal{L}(N_1)\,dx & \int_a^b N_n\mathcal{L}(N_2)\,dx & \cdots & \int_a^b N_n\mathcal{L}(N_n)\,dx \end{bmatrix} \begin{bmatrix} c_1 \\ c_2 \\ c_3 \\ c_4 \end{bmatrix} = -\begin{bmatrix} \int_a^b N_1 f\,dx \\ \int_a^b N_2\mathcal{L}f\,dx \\ \vdots \\ \int_a^b N_n f\,dx \end{bmatrix} \quad (11.23)$$

The following points concerning the Galerkin method should be noted:

- The Galerkin method creates a symmetric positive definite coefficient matrix if the differential operator is self-adjoint.
- The Galerkin approach is computationally more efficient than the least square method.

Example 11.4 Trigonometric approximation to deflection curve. Yet again we consider the now familiar beam deflection problem. The trigonometric trial function is given by

$$u(x) = A\sin\frac{\pi x}{L} = c_1 N_1$$

Where $N_1 = \sin\frac{\pi x}{L}$ and the residual

$$R(x, A) = -EI\frac{A\pi^2}{L^2}\sin\frac{\pi x}{L} - M_0$$

Solution. The undetermined coefficients in the approximate solution are obtained by equating the integral over the interval $[0, L]$ of the weighted residual to zero. The equation for the weighted residual, as expressed in equation (11.22), can be reformulated as follows:

$$\int_0^L N_1 R(x)dx = \int_0^L \sin\frac{\pi x}{L}\left(-EI\frac{A\pi^2}{L^2}\sin\frac{\pi x}{L} - M_0\right)dx = 0$$

The integration yields the following equation

$$\left(-EI\frac{\pi^2}{L^2}\right)A + \frac{2M_0L}{\pi} = 0\,.$$

Solving for A, we have

$$A = -\frac{4M_0L^2}{\pi^3EI}$$

And thus, the approximate solution is

$$u(x) = -\frac{4M_0L^2}{\pi^3EI}\sin\frac{\pi x}{L}\,.$$

Figure 11-2 shows a strong agreement between the approximate solution u(x) and the exact solution y(x) across the interval $[0, L]$. Furthermore, the approximate solution seems to be significantly more precise than the answer obtained by the point collocation approach.

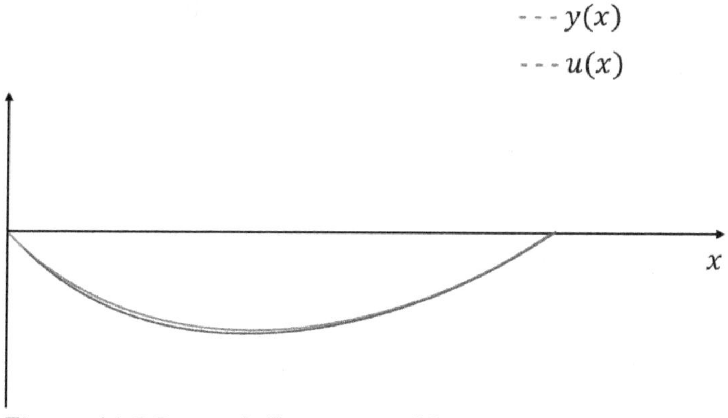

Figure 11.2 Beam deflection problem– result of Galerkin method

Polynomial approximation of the deflection curve: We use the trial function of a second-degree polynomial

$$u(x) = cx(x - L) = c_1N_1$$

where $N_1 = x(x - L)$ and the residual

$$R(x, c) = EI \times 2c - M_0$$

Integrating the weighted residual over $[0, L]$

$$\int_0^L N_1 R(x) dx = \int_0^L x(x - L)\,(2EIc2c - M_0) dx = 0$$

to obtain the following equation

$$(2EIc - M_0)\left(\frac{L^3}{3} - \frac{L^2}{2}\right) = 0$$

Solving for c, we have

$$c = \frac{M_0}{2EI}$$

thus, the approximate solution is

$$u(x) = -\frac{M_0}{2EI} x(L - x).$$

Selecting a second-order polynomial as an approximation function provides an exact solution, just like with previous approaches.

Example 11.5 Solve the differential equation

$$\frac{d^2y}{dx^2} + y = x, \quad x \in [0,2]$$

with the boundary conditions $y(0) = 0,\ y(2) = 5$ using Galerkin method.

Solution. We use the same trial function $u(x)$ as with the point collocation method:

$u(x) = 2.5x + c_2 x(x - 2) + c_3 x^2(x - 2) = 2.5N_1 + c_2 N_2 + c_3 N_3$ so that $N_2 = x(x - 2)$ and $N_3 = x^2(x - 2)$. The residual of the differential equation is given by

$$R(x) = \frac{d^2u}{dx^2} + u - x\,.$$

After duly substituting u and u'' in the above residual equation, we get

$$R(x) = 2c_2 + c_3(6x - 4) + 2.5x + c_2x(x - 2) + c_2x^2(x - 2) - x\,.$$

The unknown coefficients in the approximate solution are determined by using equation (11.22):

$$\int_0^2 x(x - 2)R(x)dx = 0$$
$$\int_0^2 x^2(x - 2)R(x)dx = 0$$

which gives the two algebraic equations for c_2 and c_3:

$$4c_2 + 4c_3 = -5$$
$$2c_2 + 4c_3 = -3$$

Solving the above set of equations for c_2 and c_3 and substitute in the assumed trial function to obtain

$$\begin{aligned} u(x) &= \left(\frac{5}{2}\right)x - x(x - 2) - \left(\frac{1}{4}\right)x^2(x - 2) \\ &= -\left(\frac{1}{4}\right)x^3 - \left(\frac{1}{2}\right)x^2 + \left(\frac{9}{2}\right)x\,. \end{aligned}$$

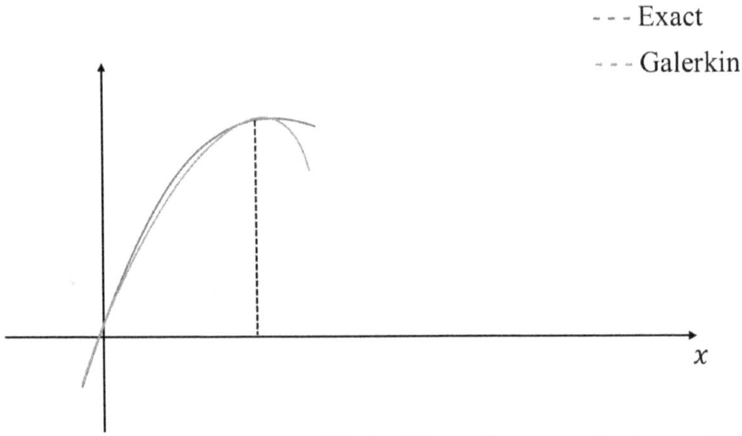

Figure 11.3 Comparison of Galerkin and exact solutions of problem 11.5

Figure 11.3 demonstrates that the approximate solution $u(x)$ closely aligns with the exact solution $y(x)$ on the interval $[0, 2]$.

We applied various weighted residual methods to solve boundary value problems. It is evident that when addressing the beam deflection problem, all methods will produce the precise solution if an approximating function of at least second degree polynomial is used. The reason for this is that the deflection curve follows a parabolic shape, specifically a second-degree polynomial. The choice of a sinusoidal function as an approximating function leads to varying solutions across different methods, with the exception of the least square and Galerkin methods. Now, one may wonder which method yields the most precise outcomes. Regrettably, a definitive answer for this remains elusive. The error is contingent upon the approximating function and the differential equation that needs to be solved. In most cases, the Galerkin method yields the most optimal outcomes.

11.5 Finite Element Method

The Rayleigh Ritz and Galerkin methods, as previously discussed, are not suitable for directly obtaining global approximate solutions for engineering problems. One significant factor contributing to this is the challenge of selecting trial functions that meet the boundary conditions, especially for complex boundaries. This indicates that the application is limited to problems with a straightforward geometry. Additionally, the use of high-order polynomials is necessary to achieve accurate global solutions. The finite element method combines the principles of the Rayleigh-Ritz and Galerkin methods to address the challenges mentioned above.

The finite element method involves dividing the region of interest into smaller subregions, known as elements. Over each element, a variational formulation of the given differential equation is created using simple functions for approximations. The individual components are subsequently combined, and the equations for the entire problem are derived through a systematic application of the variational method. In order to achieve greater accuracy, there is no need to increase the order of the functions employed. Instead, using a more refined mesh would suffice. This approach overcomes the challenges faced when directly applying variational methods. Here is a brief overview of the steps involved in the finite element method:

(i) Discretization: The given domain is divided into a number of finite elements through a process called discretization. The points where the lines intersect are referred to as nodes. Both the nodes and the elements are assigned numerical identifiers.
(ii) Derivation of element equations: A variational formulation is constructed over a typical element for the given differential equation. The element equations are derived by substituting a representative dependent variable, for example

$$u = \sum_{i=1}^{n} u_i \psi_i$$

Formulating the problem in a concise and academic manner, we delve into the variational formulation. After selecting ψ_i, the interpolation functions, the element matrices are computed.
(iii) Assembly: The subsequent step involves assembling the element equations to ensure a seamless and coherent overall solution. After completing the necessary steps, the entire system can be represented in matrix form as $Ku' = F'$. Here, K represents the assemblage property matrix, while F consists of column vectors containing unknowns and external forces.
(iv) Boundary conditions: The system of equations is adjusted by incorporating the relevant boundary conditions of the problem.

(v) Solution of the equations: Once the boundary conditions are taken into account, the system can be solved using a standard technique such as LU decomposition. The preceding steps are commonly found in most finite element approaches, making them widely applicable. In the upcoming section, the steps will be detailed and illustrated using an example of a one-dimensional problem. Due to the complexity of finite element analysis for two-dimensional problems modelled by partial differential equations, they are not discussed in this context.

11.6 Finite-Element Application in One Dimension

One-dimensional form of Poisson's equation

$$\frac{d^2T}{dx^2} = -f(x) \tag{11.24}$$

where $f(x) = a$ function defining a heat source along the rod and where the ends of the rod are held at fixed temperatures,

$$T(0,t) = T_1$$

And

$$T(L,t) = T_2$$

It is important to note that this equation is not a partial differential equation, but rather a boundary-value ordinary differential equation. This model is chosen for its simplicity, as it enables us to introduce the finite-element approach while avoiding the complexities associated with a two-dimensional PDE, for instance.

Example 11.5 Solve Eq. (11.24) for a 10-cm rod with boundary conditions of $T(0,t) = 40$ and $T(10,t) = 200$ and a uniform heat source of $f(x) = 10$.

Solution. The equation to be solved is

$$\frac{d^2T}{dx^2} = -10$$

Assume a solution of the form

$$T = ax^2 + bx + c$$

which can be differentiated twice to give $T'' = 2a$. Substituting this result into the differential equation gives $a = -5$. The boundary conditions can be used to evaluate the remaining coefficients. For the first condition at $x = 0$,

$$40 = -5(0)^2 + \mathrm{b}(0) + \mathrm{c}$$

Or $c = 40$. Similarly, for the second condition,

$$200 = -5(10)^2 + \mathrm{b}(10) + 40$$

which can be solved for $b = 66$. Therefore, the final solution is

$$T = -5x^2 + 66x + 40$$

11.7 Application to Two-Dimensional Problems

Applying the Rayleigh-Ritz and Galerkin methods to two-dimensional problems can be more complex due to the increased number of parameters that need to be determined. Here, we provide an example to demonstrate the application of the Ritz method.

Example 11.6 We consider Poisson's equation

$$\frac{\partial^2 y}{\partial x^2} + \frac{\partial^2 y}{\partial y^2} k\,, \qquad 0 \le x, y \le 1 \tag{i}$$

with $u = 0$ on the boundary C of the region S. The functional for the above problem is given by

$$I(v) = \iint\limits_S v\left(2k - \frac{\partial^2 v}{\partial x^2} - \frac{\partial^2 v}{\partial y^2}\right) dxdy \tag{ii}$$

where v vanishes on the boundary C. Let

$$v(x, y) = \alpha xy(x-1)(y-1) \tag{iii}$$

be a first approximation to u. Clearly, v satisfies the boundary conditions, i.e. $v = 0$ on the boundary C. The derivatives are given by

$$\left.\begin{aligned} \frac{\partial v}{\partial x} &= \alpha y(y-1)(2x-1); \quad \frac{\partial v}{\partial y} = \alpha x(x-1)(2y-1); \\ \frac{\partial^2 v}{\partial x^2} &= 2\alpha y(y-1); \quad \frac{\partial^2 v}{\partial y^2} = 2\alpha x(x-1)\,. \end{aligned}\right\} \tag{iv}$$

Substituting for v in Eq. (ii), we obtain

$$I(v) = \int_0^1 \int_0^1 \alpha xy(x-1)\,(y-1)[2k - 2\alpha y(y-1) - 2\alpha x(x-1)]dxdy. \tag{v}$$

Let

$$\left.\begin{aligned} a &= \int_0^1 \int_0^1 xy(x-1)\,(y-1)dxdy = \frac{1}{36} \\ a &= \int_0^1 \int_0^1 xy^2(x-1)\,(y-1)^2 dxdy = -\frac{1}{180} \\ a &= \int_0^1 \int_0^1 x^2 y(x-1)^2\,(y-1)dxdy = -\frac{1}{180} \end{aligned}\right\} \tag{vi}$$

Equation (v) now simplifies to

$$I(v) = 2k\alpha a - 2\alpha^2 b - 2\alpha^2 c$$

Hence,

$$\frac{\partial I}{\partial \alpha} = 0 = 2ka - 4\alpha b - 4\alpha c$$

Thus

$$\alpha = {}^{ak}\!/_{2(b+c)} = -{}^{5}\!/_{4}k\,, \text{ using (vi).}$$

It follows that the required approximation for u is given by

$$u \approx v = -\frac{5}{4kxy(x-1)(y-1)}.$$

The student should verify that the Galerkin method gives the same solution as above.

11.8 Exercises

1. Explain the difference between a function and a functional. Prove that
(a) $\delta(dx) = d(\delta x)$.
(b) $\delta \int f(x)\,dx = \int \delta f(x) dx$.

2.
$$\frac{\partial c}{\partial t} = D\frac{\partial^2 c}{\partial x^2} - U\frac{\partial c}{\partial x} - kc$$
where c =concentration (mg/m3), t =time (min), D =a diffusion coefficient (m2/min), x =distance along the tank's longitudinal axis (m) where $x = 0$ at the tank's inlet, U =velocity in the x direction (m/min), and k =a reaction rate (min21) whereby the chemical decays to another form. Use Galerkin's method to develop an element equation for a steady-state version of the advection-diffusion equation.

3. Obtain functionals for the boundary value problem
$$\frac{d^2y}{dx^2} = g(x), \quad y(0) = y(1) = 0\,.$$

4. Obtain functionals for the boundary value problem
$$\frac{d^2y}{dx^2} + ky = x^3, \quad y(a) = 0, \left[\frac{dy}{dx}\right]_{x=b} = 1\,.$$

5. Obtain functionals for the boundary value problem
$$x^2\frac{d^2y}{dx^2} + 2x\frac{dy}{dx} = g(x), \quad y(0) = y(1) = 0\,.$$

6. Solve the boundary value problem by Rayleigh-Ritz method
$$\frac{d^2y}{dx^2} + y = x^2, \qquad y(0) = y(1) = 0$$

7. Solve the boundary value problem by Rayleigh-Ritz method
$$\frac{d^2y}{dx^2} + 2x = 0, \qquad y(0) = y(1) = 0$$

8. Solve the boundary value problem by Rayleigh-Ritz method

$$\frac{d^2y}{dx^2} - 64y + 10 = 0, \qquad y(0) = y(1) = 0$$

9. Apply Galerkin's method to solve the boundary value problem

$$\frac{d^2y}{dx^2} + y = x^2, \qquad y(0) = y(1) = 0$$

10. Apply Galerkin's method to solve the boundary value problem

$$\frac{d^2y}{dx^2} - 64y + 10 = 0, \qquad y(0) = y(1) = 0$$

11. Use the Rayleigh-Ritz method to solve the beam deflection problem given by the variational principle

$$\prod[y] = \int_0^L \left[\frac{EI}{2}\left(\frac{dy}{dx}\right)^2 + M_0 y\right] dx$$

With the boundary conditions $y(0) = 0 = y(L)$.

12. Consider the following linear boundary value problem governed by the following generic second order linear ordinary differential equation:

$$\frac{d^2y}{dx^2} + Q(x)y = F(x), \qquad x \in [0,1]$$

With the boundary conditions $y(0) = 0,\ y(1) = Y$.
Use Galerkin method to solve the above boundary value problem.

13. Establish the Euler-Lagrange equation

$$\frac{\partial f}{\partial y} - \frac{d}{dx}\left(\frac{\partial f}{\partial y'}\right) = 0\,,$$

For a functional defined by

$$I(y) = \int_a^b f(x, y, y')dx\,.$$

CHAPTER 12

Introduction to Spectral Methods in Numerical Analysis

12.1 Introduction

Spectral approaches are one of several techniques for digitally representing a function. The fundamental concept behind numerical approaches is to estimate any function using polynomials, as they are the only functions that a computer can precisely calculate. In this context, the function u can be approximated by,

$$\hat{u} = \sum_{n=0}^{N} \hat{u}_n \Phi_n$$

where the Φ_n represent polynomials known as the trial functions. Different classes of numerical approaches can be generated based on the selection of trial functions. Finite difference schemes are derived by selecting local polynomials of a low degree, as an illustration. In the following discussion, we will focus on spectral approaches that utilise global polynomials, such as Legendre or Chebyshev polynomials. Spectral approaches, being more advanced than finite difference schemes, have demonstrated their capability to address a diverse range of issues over an extended period of time. Specifically, they enable the achievement of high levels of accuracy using little computer resources.

12.2 Solving Two-Point Boundary Value Problems using Spectral Methods

In We begin by summarizing a number of results about Chebyshev polynomials.

Definition: $T_n(x) = \cos(n \arccos x)$ (12.1)

Orthogonality relation:

$$\int_{-1}^{1} \frac{T_i(x)T_j(x)}{\sqrt{1-x^2}} dx = 0 \ , \ \ i \neq j \ ,$$

Explicit form of the first five polynomials:

$$\begin{gathered} T_0(x) = 1\,, \\ T_1(x) = x\,, \\ T_2(x) = 2x^2 - 1 \\ T_3(x) = 4x^3 - 3x \\ T_4(x) = 8x^4 - 8x^2 + 1 \end{gathered}$$

Three-term recurrence relation:

$$T_{n+1}(x) = 2xT_n(x) - T_{n-1}(x) \tag{12.2}$$

Consider the problem of finding a function $u(x)$ such that

$$-u'' + u = f(x), \quad x \in [0,1] \tag{12.3}$$

$$u(-1) = 0\,, \tag{12.4}$$

$$u(1) = 0\,. \tag{12.5}$$

The adjustment from the interval $(0,1)$ to $(-1,1)$ is made in order to fit the unique characteristics of the Chebyshev polynomials, which are specified on the interval $(-1,1)$.

We look for an approximate solution by employing a certain format.

$$u_N(x) = \sum_{i=1}^{N} v_i T_{i-1}(x). \tag{12.6}$$

The adjustment made to the index of the Chebyshev polynomial is necessary in order to incorporate $T_0(x) = 1$ into our basis. To proceed in the most straightforward manner, we can require that u^ solves the differential equation (12.3) at $N - 2$ collocation points $\{\zeta_i\}$, while also ensuring that $u_N(x)$ satisfies the boundary requirements

$$u_N(-1) = \sum_{i=1}^{N} v_i T_{i-1}(-1) = 0$$

And

$$u_N(1) = \sum_{i=1}^{N} v_i T_{i-1}(1) = 0$$

This will lead to a system of N equations in N unknowns.
Let's illustrate this by working through two specific examples in some detail.

Example 12.1 First we will take perhaps the smallest possible case, $N = 3$. We therefore are looking for an approximation in the form

$$u_3(x) = v_1 T_0(x) + v_2 T_1(x) + v_3 T_2(x) .$$

Applying the boundary conditions gives two equations:

$$v_1 T_0(-1) + v_2 T_1(-1) + v_3 T_2(-1) = 0 ,$$
$$v_1 T_0(1) + v_2 T_1(1) + v_3 T_2(1) = 0 .$$

We get a third equation by imposing the differential equation at an interior point. The obvious choice seems to be x = 0, because that is the middle of the interval. We thus have

$$v_1 \left(-\frac{d^2 T_0}{dx^2} + T_0 \right)(0) + v_2 \left(-\frac{d^2 T_1}{dx^2} + T_1 \right)(0) + v_3 \left(-\frac{d^2 T_2}{dx^2} + T_2 \right)(0) = 1 .$$

The Chebyshev polynomials are simple, making it straightforward to solve the $3x3$ matrix problem:

$$\begin{bmatrix} 1 & -1 & 1 \\ 1 & 0 & -5 \\ 1 & 1 & 1 \end{bmatrix} \begin{bmatrix} v_1 \\ v_2 \\ v_3 \end{bmatrix} = \begin{bmatrix} 0 \\ 1 \\ 0 \end{bmatrix} .$$

It is important to observe that we have arranged this in a manner where the left endpoint ($x = -1$) is positioned as the top row, the right endpoint ($x = 1$) is positioned as the bottom row, and the inner point ($x = 0$) is positioned as the centre row. The system can be readily solved to get,

$$v_1 = \frac{1}{6}, \qquad v_2 = 0, \qquad v_3 = -\frac{1}{6}.$$

The exact solution is

$$u(x) = 1 - \beta e^{x} - \beta e^{-x}, \qquad \beta = \frac{e}{e^2 + 1}.$$

Despite using only three functions for our computation, we obtained a satisfactory level of accuracy when compared to the finite difference and finite element cases.

Example 12.2 Now let's consider the same problem, but this time we will take $N = 5$ terms in our approximation. We are therefore looking for the coefficients in the expansion

$$u_5(x) = v_1 T_0(x) + v_2 T_1(x) + v_3 T_2(x) + v_4 T_3(x) + v_5 T_4(x).$$

The process remains unchanged from before. By applying the boundary conditions, we get at the two equations

$$\begin{gathered} v_1 T_0(-1) + v_2 T_1(-1) + v_3 T_2(-1) + v_4 T_3(-1) + v_5 T_4(-1) = 0, \\ v_1 T_0(1) + v_2 T_1(1) + v_3 T_2(1) + v_4 T_3(1) + v_5 T_4(1) = 0, \end{gathered}$$

which become

$$\begin{aligned} v_1 - v_2 + v_3 - v_4 + v_5 &= 0, \\ v_1 + v_2 + v_3 + v_4 + v_5 &= 0. \end{aligned}$$

We now need to apply our approximation at three interior collocation points. It might seem obvious to take the middle three from the equidistant set of points

$$\zeta_k^{(5)} = \left\{-1, -\frac{1}{2}, 0, \frac{1}{2}, 1\right\},$$

but it is better to use the so-called Gauss-Lobatto Chebyshev points defined *by*$_4$

$$\zeta_k^{(5)} = \cos\frac{\pi k}{N-1}. \tag{12.7}$$

We will work this example with both sets of collocation points. Regardless of our choice of points, we need to construct and solve the 5×5 system

$$\begin{bmatrix} 1 & -1 & 1 & -1 & 1 \\ 1 & a_{22} & a_{23} & a_{24} & a_{25} \\ 1 & a_{32} & a_{33} & a_{34} & a_{35} \\ 1 & a_{42} & a_{43} & a_{44} & a_{45} \\ 1 & 1 & 1 & 1 & 1 \end{bmatrix}, \begin{bmatrix} v_1 \\ v_2 \\ v_3 \\ v_4 \\ v_5 \end{bmatrix} = \begin{bmatrix} 0 \\ 1 \\ 1 \\ 1 \\ 0 \end{bmatrix} \tag{12.8}$$

where the "interior" coefficients are defined by

$$a_{kj} = \left(- - \frac{d^2 T_{j-1}}{dx^2} + T_{j-1}\right)\left(\zeta_k^{(5)}\right), \tag{12.9}$$

For $k = 1,2,3,4$.

Case 1: *Equidistant points.* Here $\zeta_k^{(5)} = \left\{-1, -\frac{1}{2}, 0, \frac{1}{2}, 1\right\}$, and a little work with (12.8) and (12.9) shows that the matrix is

$$A = \begin{bmatrix} 1 & -1 & 1 & -1 & 1 \\ 1 & -0.5 & -4.5 & 13 & -8.5 \\ 1 & 0 & -5 & 0 & 17 \\ 1 & 0.5 & -4.5 & -13 & -8.5 \\ 1 & 1 & 1 & 1 & 1 \end{bmatrix}.$$

Figures 12.1 and 12.2 display the solution and error obtained. It should be noted that the solutions, whether approximate or exact, cannot be distinguished on the graph. The maximum absolute error is 3.0949×10^{-4}.

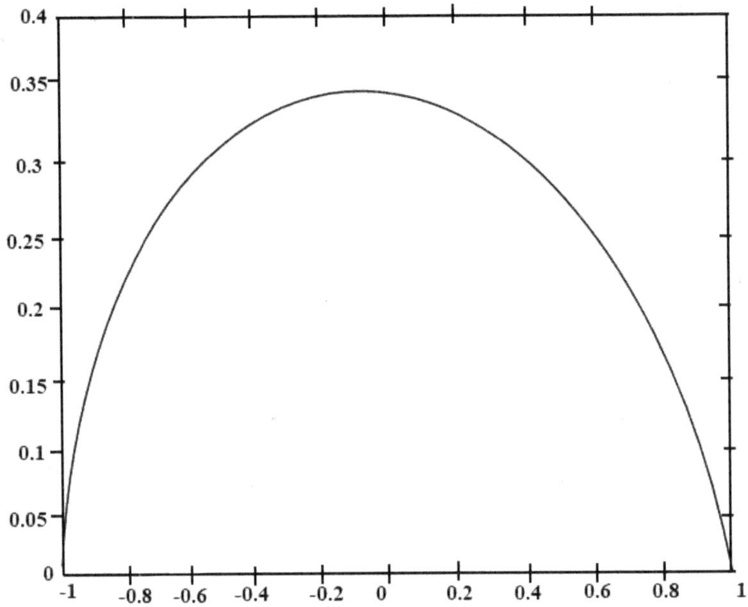

Figure 0.1 Spectral solution for a uniform grid, $N = 5$.

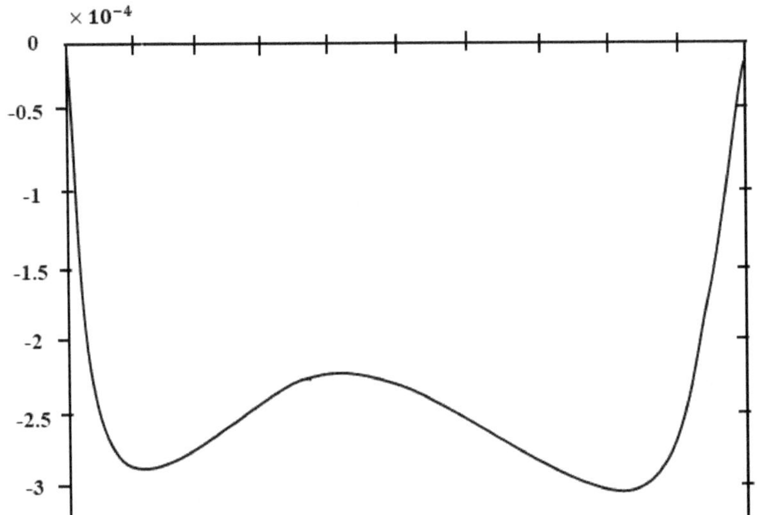

Figure 0.2 Spectral error for a uniform grid, $N = 5$.

Case 2: *Chebyshev points.* We have

$$\zeta_k^{(5)} = \{-1, -0.7071, 0, 0.7071, 1\},$$

and, again, a little work with (12.8) and (12.9) shows that the matrix now is

$$A = \begin{bmatrix} 1 & -1 & 1 & -1 & 1 \\ 1 & -0.7071 & -4 & 17.6777 & -33.0000 \\ 1 & 0 & -5 & 0 & 17.0000 \\ 1 & 0.7071 & -4 & -17.6777 & -33.0000 \\ 1 & 1 & 1 & 1 & 1 \end{bmatrix}.$$

The solution coefficients are

$$v = (0.1796, 0.0000, -0.1761, 0.0000, -0.0035)^T$$

The resulting solution and error are plotted in Figs. 12.5 and 12.6, and the approximate and exact solutions are again indistinguishable on the graph. The maximum absolute error is 1.6849×10^{-4}, which is slightly better than the uniform grid. (Note that the coefficients of the odd Chebyshev polynomials are zero.)

There is a certain aspect of this comparison that may be considered as comparing apples and oranges, as the problems are presented on different intervals and use different discretisation parameters. However, it is important to highlight that we achieved a similar level of accuracy, and in fact slightly better, with a reduced number of computational elements. By including additional terms in the spectral approximations, we can observe a rapid convergence, as we will demonstrate later on.

Nevertheless, there are certain matters that need to be discussed. It is worth noting that the matrices used in spectral methods are full, whereas we should remember that the matrices used in FD and FEM methods were tridiagonal. Upon examining the two 5×5 matrices in Example 12.2, it becomes apparent that discerning any significant structure is challenging. Furthermore, our construction of the matrices was, to put it bluntly, quite inefficient. However, there is a significant amount of organisation within these matrices (although we won't delve into it extensively in our discussion), and we will soon establish a more comprehensive and systematic approach to creating them. One advantage of using full matrices is that we can achieve high accuracy without needing a matrix as large as the ones required for FD or FEM approximations.

12.3 Solving Time-Dependent Problems using Spectral Methods

A common approach for solving time-dependent problems with spectral methods includes using a spectral approximation for the spatial variables. This simplifies the original problem into a system of ordinary differential equations (ODEs). The "method of lines" approach enables us to apply any ordinary differential equation method to obtain the final approximation. We ensure precise spectral accuracy in the spatial variables, while relying solely on the ODE method for time variation accuracy. This is considered acceptable due to the cost-effectiveness of taking smaller time steps compared to refining the spatial approximation.

To provide a more focused illustration, let us examine the following example problem:

$$u_t = u_{xx}, -1 \leq x \leq 1 \tag{12.10}$$

$$u(-1,t) = u(1,t) = 0,\ t > 0, \tag{12.11}$$

$$u(x,0) = \cos \pi x/2 - \sin \pi x\,, \tag{12.12}$$

which has the exact solution $u(x,t) = e^{-\left(\frac{\pi}{2}\right)^2} t \cos \frac{\pi x}{2} - e^{-\pi^2} t \sin \pi x$. Let's look for an approximation in the form

$$u_N(x,t) = \sum_{k=1}^{N-1} v_k(t) C_{k+1}(x)\,,$$

The C_i values are defined as in equations

$$C_{2n}(x) = T_{2n}(x) - T_0(x)$$
$$C_{2n+1}(x) = T_{2n+1}(x) - T_1(x)$$

By substituting this into the given PDE (12.10) and evaluating it at the collocation points $\left\{\zeta_j^{(N)}\right\}_{j=1}^{j=N-2}$, we can promptly obtain the resulting equation

$$\sum_{k=1}^{N-1} v'_k(t)C_{k+1}\left(\zeta_j^{(N)}\right) = \left(\sum_{k=1}^{N-1} v_k(t)C''_{k+1}\left(\zeta_j^{(N)}\right)\right), 1 \leq k \leq N-1,$$
$$1 \leq j \leq N-2.$$

The boundary conditions are fulfilled automatically, leading to the following system of ordinary differential equations (ODEs) for the coefficients $v_k(t)$:

$$Mv'(t) = Kv(t),\ v(0) = v_0, \tag{12.13}$$

where the matrices M and K are defined by

$$M = \begin{bmatrix} C_2\left(\zeta_1^{(N)}\right) & C_3\left(\zeta_1^{(N)}\right) & \cdots & \cdots & C_{N+1}\left(\zeta_1^{(N)}\right) \\ C_2\left(\zeta_2^{(N)}\right) & \vdots & \vdots & \vdots & C_{N+1}\left(\zeta_2^{(N)}\right) \\ \vdots & \vdots & \vdots & \vdots & \vdots \\ C_2\left(\zeta_{N-2}^{(N)}\right) & \cdots & \cdots & \cdots & C_{N+1}\left(\zeta_{N-2}^{(N)}\right) \end{bmatrix},$$

And

$$k = \begin{bmatrix} C''_2\left(\zeta_1^{(N)}\right) & C''_3\left(\zeta_1^{(N)}\right) & \cdots & \cdots & C''_{N+1}\left(\zeta_1^{(N)}\right) \\ C''_2\left(\zeta_2^{(N)}\right) & \vdots & \vdots & \vdots & C''_{N+1}\left(\zeta_2^{(N)}\right) \\ \vdots & \vdots & \vdots & \vdots & \vdots \\ C''_2\left(\zeta_{N-2}^{(N)}\right) & \cdots & \cdots & \cdots & C''_{N+1}\left(\zeta_{N-2}^{(N)}\right) \end{bmatrix}.$$

The vector $v(t)$ represents the coefficients in our spectral expansion.

$$v(t) = \left(v_1(t), v_2(t), \ldots, v_{N-2}(t)\right)^T,$$

and the initial condition v_0 is defined by a collocation approximation to the initial function, $v_0(x)$; i.e., we solve the linear system

$$Mw = \left(u_0\left(\zeta_1^{(N)}\right), u_0\left(\zeta_2^{(N)}\right), \ldots, u_0\left(\zeta_{N-2}^{(N)}\right)\right)^T$$

And set

$$v_0 = \sum_{k=1}^{N} w_k C_{K+1}(x)\,.$$

To finish the approximation, we need only decide on a method to solve the ODE system (12.13). We will begin by utilising the trapezoid rule method. This method is based on the Crank-Nicolson idea, allowing us to

$$Mv_{n+1} = Mv_n + \frac{1}{2}\Delta t(Kv_n + Kv_{n+1})\,,$$

Or,

$$\left(M - \frac{1}{2}\Delta tK\right)v_{n+1} = \left(M - \frac{1}{2}\Delta tK\right)v_n\,, \qquad n = 1,2,\ldots.$$

In order to improve error performance over time, it is necessary to apply a reduced time step or a more precise ODE method. Now, let's look at some of the different possibilities for the second option

- Fourth-order Runge-Kutta
- Higher-order Adams-Bashforth methods
- Higher-order Adams-Moulton methods
- Higher-order BDF (Backward Differentiation Formula) methods

Each of these options has the potential to be effective, each with its own set of advantages and disadvantages.
However, due to concerns regarding the stability of the explicit finite difference method for the heat equation, we are cautious about considering the first two options, as they are also explicit methods. We will now examine the final two options, which are implicit and therefore more stable compared to the explicit methods.

Example 12.3 Adams-Moulton

As we have a second-order method—the trapezoid rule/Crank-Nicolson—which is also a second-order Adams-Moulton method—we will examine the third-order Adams-Moulton method, which is typically denoted as

$$y_{n+1} = y_n + \frac{h}{12}(5f_{n+1} + 8f_n - f_{n-1})\,,$$

where $y' = f(t, y)$ is the differential equation. Our differential equation is given by (12.13), and is a system of equations in which $f(t, y) \leftrightarrow M^{-1}Kv$. So,

$$v_{n+1} = v_n + \frac{\Delta t}{12}(5M^{-1}Kv_{n+1} + 8M^{-1}Kv_n - M^{-1}Kv_{n-1}).$$

Simplifying this, we get

$$\left(M - \frac{5\Delta t}{12}K\right)v_{n+1} = \left(M + \frac{8\Delta t}{12}K\right)v_n - Kv_{n-1}. \tag{12.14}$$

A single starting value, v_1, can be obtained using the trapezoid rule method for a single step. From there, it is a simple recursion to generate new solution values from (12.14). When we apply this to our example problem with $N = 8,\ \Delta t = \frac{1}{N}, T_{\max} = 2.5$, and $T_{\max} = 2.5$, the resulting plot can be seen in Figure 12.3. It is worth noting that the maximum error starts to increase for $t > 1$ (approximately). This appears to be negative. It is important to note that multistep methods, such as Adams-Moulton, can encounter stability problems. In order to ensure the stability of the computation, it is necessary to use a sufficiently small step size.

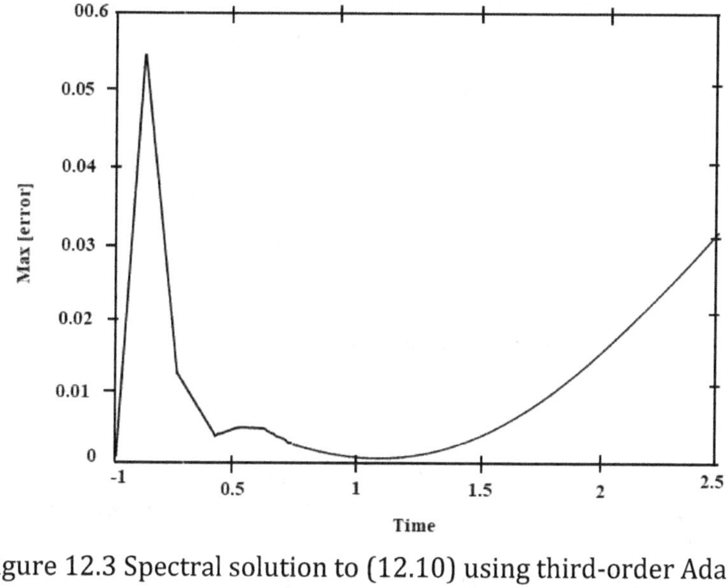

Figure 12.3 Spectral solution to (12.10) using third-order Adams Moulton time-stepping

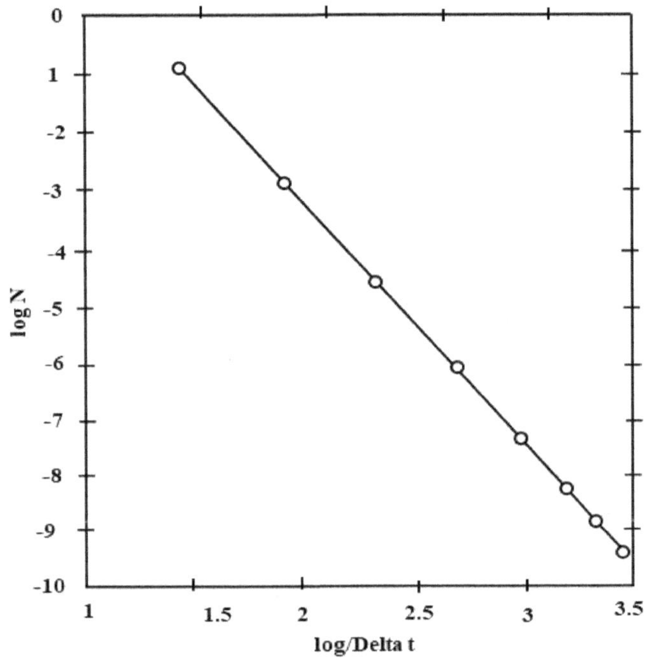

Figure 12.4 Plot (log-log) of largest stable Δt vs. N

We must choose Δt so that $\Delta t \lambda_k$ is within the stability region for all eigenvalues λ_k; of the matrix $G = M^{-1}K$. For our particular case here, we have that all the eigenvalues are real, with $\max_k |\lambda_k| = 131$ and $\Delta t = \frac{1}{8}$, so the product $\Delta t \lambda_k = 16.375$ is clearly not inside the stability region. We need to have $\Delta t \leq 6/\lambda_{\max}$; thus, $\Delta t = 0.458 = c/N$ for $c = 0.3664$ and now $N = 8$ will suffice. If we now try $\Delta t = 0.3664/N$ for the $N = 12$ case, we find that the computation blows up catastrophically. The maximum eigenvalue for $M^{-1}K$ is now over 700— the eigenvalues are growing too fast for our choice of time steps. Fig 12-4 shows a log-log plot of the largest possible At as a function of N, and a least squares analysis suggests that a relationship of the form

$$\Delta t = CN^{-4}$$

is present. Clearly, the optimal time step we can utilise will decrease rapidly. In this situation, we are faced with two options: either we can continue using very small time steps, or we can explore the possibility of employing a more stable time-stepping method, such as the BDF (Backward Differentiation Formula) family.

Example 12.3 BDF

The BDF methods are stable along the entire negative real axis, so they appear to be the appropriate methods to use. The third-order BDF formula is

$$y_{n+1} - \frac{6h}{11} f(t_{n+1}, y_{n+1}) = \frac{18}{11} y_n - \frac{9}{11} y_{n-1} + \frac{2}{11} y_{n-2}$$

and the fourth-order formula is

$$y_{n+1} - \frac{12h}{25} f(t_{n+1}, y_{n+1}) = \frac{48}{25} y_n - \frac{36}{25} y_{n-1} + \frac{16}{25} y_{n-2} - \frac{3}{25} y_{n-3}$$

Applied to our particular problem (recall $f \leftrightarrow M^{-1}K$ we have

$$Mv_{n+1} - \frac{6h}{11} K v_{n+1} = M\left(\frac{18}{11} v_n - \frac{9}{11} v_{n-1} + \frac{2}{11} v_{n-2}\right),$$

$$Mv_{n+1} - \frac{12h}{25} K v_{n+1} = M\left(\frac{48}{25} v_n - \frac{36}{25} v_{n-1} + \frac{16}{25} v_{n-2} - \frac{3}{25} v_{n-3}\right),$$

We will start with the third-order BDF method, using our second-order Crank Nicolson code to generate the two necessary starting values. Table 12.1 shows the maximum absolute error over time. Two things are of concern or interest here: (1) The error does not appear to be going down by the expected factor of 8 that we should expect from an $\mathcal{O}(\Delta t^3)$ method (in fact, a least squares fit to the logarithms of the data suggests that the order is around 2); (2) In all four cases, the maximum absolute error occurred at the second or third step, i.e., at one of the starting values. Perhaps using an $\mathcal{O}(\Delta t^2)$ method to generate the starting values is not such a good idea in this case.

Table 12.1 Maximum error for BDF3 time-stepping using Crank-Nicolson starting values

N	$\max_{t\leq 2.5}\lvert u - u_N \rvert$
4	0.12205771461309
8	0.05536342126983
16	0.01216091356321
32	0.00271680076022

If a more efficient third-order single-step method were available, it would be preferred. Implicit Runge-Kutta methods are known to be challenging to derive and come with high computational costs. Instead, we will persist in applying the Crank-Nicolson method for the initial values, but with reduced time intervals.

To achieve the desired outcome, it is important to use a sequence of brief time intervals, denoted as Δt , ensuring that ΔT (the desired step using the high-order method) is a whole number multiple of Δt. In other words, $\Delta T = M\Delta t$, where M represents an integer. Our initial value is the approximate solution at $M\Delta t$, and the second one is the approximate solution at $2M\Delta t$. We implemented this for the BDF3 scheme, using the formula $\Delta T = 0.25/N$ and $\Delta t = \Delta T/32$.

We calculated up to $t = 2.5$ and monitored two distinct errors

$$E_{\max} = \max_{0<t\leq 2.5} \|u(.,t) - u_N(.,t)\|_\infty$$

(this is the maximum pointwise error in the entire computation), and

$$E_1 = \|u(.,1) - u_N(.,1)\|_\infty$$

(this is simply the pointwise error at the arbitrary value $t = 1$). We got the results in Table 12.2. (Note that we are no longer computing the $N = 4$ case; the results are simply too crude.)

Table 12.2 Errors for BDF3

N	$E_{\max}$	E_1
8	0.00196351553612	0.00002593734915
12	0.00062352684952	0.00000761067383
16	0.00027970078950	0.00000318853246
20	0.00014831615102	0.00000162602736
24	0.00008808570623	0.00000093852185
28	0.00005651598554	0.00000058995674
32	0.00003836696856	0.00000039469439

If we do the same kind of thing using the BDF4 formula, we get the results in Table 12.3.

Table 12.3 Errors for BDF4

N	$E_{\max}$	E_1
8	0.00065906557204	0.00000351957149
12	0.00009256524086	0.00000047008815
16	0.00003163313807	0.00000014416519
20	0.00001375731857	0.00000005804163
24	0.00000691388232	0.00000002770166
28	0.00000383871479	0.00000001485340
32	0.00000230094409	0.00000000866881

It is apparent that both errors are decreasing at the expected rate.

12.4 Exercises

1. Use spectral collocation with Crank-Nicolson time-stepping to solve the following PDE:

$$u_t = u_{xx}\,,$$
$$u(-1,t) = 0\,,$$
$$u(1,t) = 0\,,$$
$$u(x,0) = \cos\frac{\pi x}{2} - \sin 4\pi x\,.$$

The exact solution is $u(x,t) = e^{-\pi^2 t/4}\cos\frac{\pi x}{2} - e^{-16\pi^2 t}\sin 4\pi x$. Compute out to $t = 1$; use a sequence of values of N ; plot your approximation and the error for one of them at $t = 1$.

2. Use your spectral code to solve the problem

$$u_t = au_{xx}\,,$$
$$u(-1,t) = 1\,,$$
$$u(1,t) = 1\,,$$
$$u(x,0) = (x^2 - 1)^8$$

Assume that $a = 1$ and compute out to $t = 1$ using a sequence of values of N. Plot the solution profile as the computation advances. Now vary a (you must keep it positive, of course) and investigate how this affects the solution.

3. Now change the initial condition to $u(x,0) = (x^4 - 1)^8$ and repeat Problem 2.

4. Consider how to implement spectral collocation with variable coefficients. Construct the general linear system that would result from solving the problem

$$u_t = au_{xx}\,,$$
$$u(-1,t) = 0\,,$$
$$u(1,t) = 0\,,$$
$$u(x,0) = u_0(x).$$

5. Apply your results from Problem 4 to approximate solutions to

$$u_t = a(x)u_{xx}\,,$$
$$u(-1,t) = 0\,,$$
$$u(1,t) = 0\,,$$
$$u(x,0) = \cos\frac{\pi x}{2}\,,$$

for the following choices of a :

a) $a(x) = (1 + x^2)$;
b) $a(x) = e^{-x^2}$;
c) $a(x) = (1 - x^2)$. (Because a vanishes at the boundary, this problem is known as degenerate, but you should be able to compute solutions.)

6. Consider the nonlinear problem

$$\begin{gathered} u_t = u_{xx} + Vuu_x \,, \\ u(-1,t) = 0 \,, \\ u(1,t) = 0 \,, \\ u(x,0) = \sin \pi x \,. \end{gathered}$$

Use spectral collocation to attack this problem by treating the non-linearity explicitly (at time $t = t_n$) and the differential equation implicitly (at time $t = t_{n+1}$).

www.ingramcontent.com/pod-product-compliance
Lightning Source LLC
LaVergne TN
LVHW052252200926
848481LV00002B/351